ALSO BY HENRY HOBHOUSE

FORCES OF CHANGE

SEEDS OF CHANGE

Henry Hobhouse

SEEDS OF WEALTH

Four Plants That Made Men Rich

Shoemaker & Hoard
Washington, DC

Library of Congress Cataloging-in-Publication Data is available.
ISBN 1-59376-044-2

Printed in the United States of America

Shoemaker & Hoard
A Division of Avalon Publishing Group Inc.
Distributed by Publishers Group West

10   9   8   7   6   5   4   3   2   1

To all those who became friends

after they had read

Seeds of Change

Author's Note

Parts of the chapter on wine appeared, in a different form, in the publication which accompanied the Columbus Quincentenery Exposition at the Smithsonian Institute, Washington DC, in 1992. The exhibition and the publication were both called *Seeds of Change.*

I acted as a consultant to the exposition and originally suggested the theme. I also wrote about wine, as well as agreeing to use the name of the book of the same title, which, of course, has a theme similar to that of the exhibition, although my book was published seven years earlier.

I was not responsible for the political correctness of the exposition. A modern blight sees history through a prism (or fog) that distorts much of the past, not because of often obvious absurdities, but because of anachronisms. Contemporary political correctness can only exist after certain conditions have been fulfilled. These conditions did not exist before current technology made them possible. So, to consider the past when these factors could not be present under the assumption that they were, is naive of students. For teachers it is at best ignorant and, at its worst, close to intellectual fraud.

In brief, Columbus was never politically correct, nor could he be, nor could anyone be whom he met in the New World.

Contents

Notes appear at the end of each chapter

Introduction

Seeds of Change, which I wrote nearly twenty years ago, was the first work that sought to give plants a role as an important causative factor in history. This was then a novel approach, somewhat imitated since. There were originally five plants and a sixth, coca, the source of cocaine and crack, was added to the paperback in 1999. Traffic in cocaine, of course, is now an enormous trade; its street value is said to be larger than the gross national product of the United Kingdom and is growing at least three times as fast.

The thrill of illegitimacy should not excite the reader of this book, which is dedicated, as the subtitle says, to plants that made men rich. The first is timber, which is important in the history of both England and the thirteen North American Colonies which became the first thirteen States. In England, timber supplies became increasingly insufficient at the time of the Reformation. As timber – then mostly used as fuel – became scarce, more and more coal was dug. Because coal had been significant as a fuel, especially in London, for several centuries before Queen Elizabeth, easily procured surface coal became deficient. Because there was little deep coal not associated with water, pumps became essential, and they became fundamental in deep mines by 1700. Though with low efficiencies, these steam-driven pumps were common in England more than fifty years before James Watt, who, we are conventionally taught, was the practical genius who made steam-power widely available, thus being responsible for the motive force behind the coal–steam–iron Industrial Revolution. In fact, it is argued, the Industrial Revolution occurred in the United Kingdom fifty to one hundred years ahead of any other country largely because of the national shortage of timber, the first such shortage in a major European power.

But timber was still needed, and it was the British–American Colonies that supplied it. The Royal and Merchant Navies (before about 1870) and the iron and steel industry (until about 1820) were almost wholly dependent upon the wealth of wood in the thirteen colonies and for more than a century before the Revolution. This dependence is demonstrated by 'mast convoys' as early as the reign of Charles I, and because even Nelson's victory at Trafalgar was won by ships largely built of American timber.

The wealth of American timber was much more significant than this; it fuelled the westward march in the 1800s and the occupation of the huge area between the Allegheny–Adirondacks and the great valley of the Mississippi. Earlier, ample timber encouraged the development of fishing and whaling and of ships both on rivers and at sea, of railroads, of wooden houses in new cities like Chicago, and timber was partly responsible for the cheap fuel philosophy which survives today, but in a different form. Houses of wood are still favoured over brick and stone, and the ability of the settler family to find everything needed in a well-timbered grant of land was essential to the American dream. Finally, because the heavy use of American timber lasted till oil and natural gas were available, there was no dominance of coal in the body politic in the later 1800s, a commonplace in England, in which country coal remained a basic fuel until the late 1980s. In contrast, until some time after the Civil War, little coal was dug in the USA.

The peculiar (and vital) difference remains that the early industrial greatness and wealth of England depended on an absence of timber while the opposite circumstance made possible an even greater form of wealth creation in the United States.

The second chapter, on wine, is about the wealth-creating potential of vines in place of more mundane crops like cereals. This has been true from before the time of Periclean Athens, through Roman and medieval experience, right up to the present. Today, in New Zealand, an entirely new wine industry has achieved world-class status within a dozen years, an amazing achievement. From the earliest times to the present, to plant vines in the right place and manage them properly, and to make wine without breaking fundamental biochemical rules, was (and is) to multiply the nett profits from the same area of land by between twenty and two hundred times. The first multiple is probably true of 'ordinary' table wine as grown and produced in France, Italy, Germany or the New World. The second, ten times as great, is true of the vineyards where fine or great wines are grown and made, in Burgundy, Bordeaux, or choice places of comparable virtue in Italy, Germany, California or Australasia.

The chapter goes further. Here is a review, technological as well as economic, of the growing and making of wine from 600 BC to date, a bold claim for an essay only 20,000 or so words long. But the truth is that familiar, ancient methods were in use as recently as shortly after the Second World War and there have been more significant changes since 1945 than there were in the previous five hundred years. No-hassle vinification, the beginnings of genetic modification, new, simple, economical techniques, the use of satellite-tracking systems to assist the gathering of the grapes, all these are combined with those shifts familiar to most farmers in the Western world – tractors in place of animal power, pruning by machine, irrigation and constant sampling, all these only made possible by technology developed elsewhere. The amount of wine that is made in the world has multiplied many times in the last fifty years. So has the (now numerous) host of men and women made wealthy because of wine. This last crowd, of course, includes those who create winespeak, a momentous, if abstruse, marketing tool.

Then there is rubber. The vital plant, *Hevea*, is one of the great gifts of Latin America to the world; but other latex-producing species were originally 'hunted' not only in the Americas, but also in Africa and India. When *Hevea* plants were successfully transferred to new plantations in the Malay Straits, supplies of natural rubber came, at first largely, then entirely from plantation *Hevea*, alone able to meet the enormous demand created by the growth of electricity, bicycles and automobiles.

Before the First World War, there was a boom on the exchanges of the world comparable with the railway mania of the 1840s or the dot.com. bubble of the late 1990s. After the War, the trade became political, and the price of rubber a bone of contention between London and Washington, in a little-known dispute. There was a slump in prices before and after the 1929 crash, and a slow pre-1941 recovery. Then there was the long agony of the Japanese occupation and the communist attempt to disrupt the world's rubber supply. For a generation before and during the Second World War, serious efforts were made to replace natural rubber derived from the *Hevea* trees with synthetics. Though these synthetic alternatives are successful, there is one unique reason why natural rubber should remain an essential for one primary purpose; and it is a fairly safe bet that not one in a thousand will guess the reason before reading the story here. Rubber has generated wealth for many, and has largely created three important new nations, in descending order of fortune: Singapore, Malaysia and Indonesia.

The last chapter is about tobacco, a plant with a very negative contemporary image but which has had historically profound effects upon Anglo-American relations. The colony of Virginia prospered because James I of England (reigned 1603–25) decreed not long after Shakespeare died, that his Old World subjects should not be allowed to grow tobacco, neither for sale nor for personal use. The ban was unique in Western Europe, every Continental being allowed to meet private needs in a private garden. King James also encouraged Virginia to save gold bullion, with which to pay for imports of 'Spanish' tobacco. Virginia was then of course a British colony, and imports from the colonies caused no drain on the gold reserves.

By 1776, tobacco in Virginia and Maryland was responsible for more than a quarter of the annual income of all thirteen Colonies. In tobacco-growing country this wealth produced a landlord class, including many of the signatories of the Declaration of Independence and outstanding leaders such as Washington, Jefferson and Madison. Little did the Stuarts, who encouraged the growth of Virginia by favouring American tobacco, ever imagine a democratic revolution resulting from their authoritarian initiative; another example of the law of unintended consequence.

Tobacco continued to be an important commodity in world trade, particularly in Anglo–American commerce, but the economic take-off only really occurred when the cigarette was invented in the 1840s, and the cigarette-making machine developed in the 1870s; marketing and advertising followed. By 1900, one firm in England and another in the United States had each become the most profitable corporation in their respective economies. Until the late

1940s, there was no respectable objective connection between the use of tobacco and cancer or heart disease and its use was considered essential to welfare in each of the two World Wars. The tobacco habit was supported in every way by wartime Allied governments. Between 1900 and 1945, cigarette use by women increased from a very few per cent to nearly 40 per cent of all adult women in the United Kingdom. Elsewhere, after 1945, particularly in Western Germany, cigarettes became a form of currency more trusted than paper money.

The story since 1950 has been one of strategic retreat by the industry, with 'safe', 'low-tar' and 'low-nicotine' cigarettes appearing in sequence. The wealth-creating factor remains, but the cost to those who sicken with cancer, heart disease or emphysema could probably be considered greater than the wealth accruing to those in the trade. But the greatest beneficiaries in most countries have been, since about 1920, the tax gatherers.

There is one unknown in the story. Is it the acidity of modern cigarettes that makes their use so dangerous? Would a return to alkaline smoke, universal before 1860, be therapeutic and reverse the causal connections with ill health?

*

There is much here about Anglo-American relations, commercial connections being as important as the politics, or more so. The critics can claim a degree of Anglocentricity. They are right. Without England, the United States would be like Mexico, and without the thirteen Colonies the United Kingdom would be a much poorer place. The wealth of which these are the seeds would have accrued elsewhere.

TIMBER

The Essential Carpet

I

There are huge areas of the world where, if humans ceased to exist, and if they had not already poisoned the Earth, trees would reclothe the land; this would not happen quickly, but it would be well under way within a century. Equally, it was often an all-pervading forest with which the Ancients were confronted, although no one now knows much about the original mix of plant species nor about the disposition of those animals originally native to the virgin habitat. At a low level of human population, forests provided an environment which men and animals could co-habit without much friction, but as human numbers grew, forests and the relationship between humans and non-humans underwent changes, often brutal.

Wood was a vital resource for the material progress of mankind – more especially as populations increased and hunter-gatherers became farmers, and even more when the earliest urban civilizations were established. By 2000 BC, ample forests were as important as land for staple foods like cereals, and one great civilization – that of Egypt – significantly had to import a great deal of its timber, notably from Phoenicia (modern Lebanon). Fortunately the Phoenicians were addicted to trade and did not excessively exploit the Egyptian weakness in wood. In other ancient countries, sufficient wood was an essential raw material for every human activity needing heat, and for housing, transport, early devices for raising water and for spinning or weaving. Even if metals were used, the

wood content of a metal artefact would probably be greater than if
the object were itself of wood. Rome, a capital city of a million
people, grossly wasted wood for several centuries and more than
partially deafforested the Mediterranean. Until the Middle Ages in
England, there was little use of coal for heating, and until the late
eighteenth century there was no way of using anything but timber
(of the right sort) to smelt metals. Few today give enough weight
to the importance of timber in history, nor is it widely known that
nearly one-third of the world's population is still dependent upon
wood for domestic fuel.

As this chapter shows, the English were the first people in
modern times to run out of home-produced wood, with far-ranging
consequences; their colonies in North America became essential,
both before and after Independence, for the Royal Navy, for
merchant ships, for iron and for the special woods for furniture
and the many manufactured objects then made of wood, like blocks
for ships' rigging, wheels of every kind, and harness.

Throughout modern history there has been an unseen conflict
in land-use between short and long term, a clash made the more
acute because population was (usually) expanding, more or less
quickly. In the short term, land-for-food was obviously more
important than land-for-timber, which equated to land-for-fuel
or land-for-buildings or land-for-ships. But what happened to a
country that could not grow its own food as well as its own timber?
The answer was that it imported whichever was cheaper, not
necessarily in money but in land-use. The Netherlands, then as
now more densely populated even than England, imported wood,
some from England before 1558, then from France, Poland and
Russia after English exports ceased. England no longer exported
wood for fuel and became short of wood for use at home. Uniquely,
the English found coal to make up the deficiency, and imported
naval stores from the Baltic and, later, from New England. England
did not import any other consumables in such considerable quantity before the nineteenth century, except for exotica like tea and
sugar, while tobacco was a special case.[1]

Of the countries geographically close to England, France had,
in 1600, seven times the land area of England but only twice the
population; the Netherlands had only 5 per cent of the area of

France but 20 per cent of the French population, or, to put it another way, 35 per cent of the area of England and 40 per cent of the population. So the Netherlands was more densely populated than England, from which timber was exported before the mid-sixteenth century, and four times more densely peopled than France, which shipped timber abroad until the eighteenth century.

Another favoured source of timber for European countries in deficit was, of course, the Baltic, the Northern part producing some of the hardest known softwoods, much prized for naval stores. Countries in the Southern Baltic, like Pomerania and Poland, produced valuable hardwoods. Trade with Russia was not highly developed until the time of Peter the Great in the early eighteenth century. In the Mediterranean, the great source of surplus timber was Turkey, and the shores of the Black Sea provided much that Europeans needed. But England, in serious deficit after 1600, was also a leading trading nation at the same date. Subject to the limitations of diplomacy and war, English timber merchants tended to buy timber wherever it was cheaper, but this was not really an alternative to substituting coal for fuel-wood. It should be remembered that the market for fuel-wood (including wood for charcoal-smelting) probably made up 90 per cent of wood used before coal became an alternative.

II

By 1666, the year of the Great Fire of London, which destroyed more than 400 acres of dense ancient building in the largest city in Western Europe, timber was already in short supply in England. When rebuilding was planned – in the middle of the Dutch War – weakness in timber resources became even more obvious. More than a million acres of mature woodland would be needed to rebuild the City as it had been. Even rebuilding in brick (and some stone), as Wren proposed, would have required more wood (as fuel) than was easily available; it was immediately clear to all that wood was critically short in England. John Evelyn's *Sylva* (1664) had already focused public attention on the problem.[2]

Shortage, as always, influenced cost. London merchants found that wood of building quality had risen in price tenfold in the century since 1560. Even domestic firewood had become so hard to find and so expensive that coal from the North-East had captured more than half the metropolitan fuel trade before Queen Elizabeth's death in 1603.

London had a population of 200,000 people in 1600, more than any other Christian city except Naples, at 300,000, more even than Paris, which had formerly had the largest population in Western Europe. In that same year Istanbul was, at 700,000, demographically second only to the largest Chinese city, Canton, probably, at 1 million, the largest city in the world. But all cities then larger than London were also naturally warmer; London was the only European city substantially heated by coal in 1600, and remained the only city (perhaps in the world) largely using coal for most of the next two centuries; it also became the first city to be lit by gas made from coal, brought by ship.

The largest use of wood in England after domestic purposes was in the iron industry, the largest single industrial consumer of fuel in the Western world from early times. In 1700, the amount of iron made worldwide cannot have been more than 300,000 tons a year, and it was universally a cottage industry, using relatively large quantities of wood, burnt as charcoal. The ratio in England in 1700 was probably more than 100 tons of wood (equivalent to less than 20 tons of charcoal) for every ton of smelted iron, and more wood was used to make wrought iron. In 1800, probably only just about 2 million tons of iron were made worldwide which needed, in turn, about 400 million tons of wood burnt as fuel. By 1900, 150 million tons of coal were burnt worldwide to produce 25 million tons of iron and steel. In 1700, it was very different. In that year, a small ironworks might have had an annual output of only 50 tons of crude iron, and this effort (at a ton a week) would have needed the services of a dozen men and a water-driven set of bellows. Each week someone would deliver 20 tons of charcoal from a radius of a few miles, the practical limit for contemporary wagons before Macadam roads.

The final load of timber, the last of what had once been more than 1,000 shiploads a year, was exported from England in 1558, and later, as wood rose faster in price than did other primary products in the inflationary age that followed the influx of Spanish-American silver, there is no mention of any wood exports. Within fifty years, the supply of silver coin had more than quadrupled in Europe, with a proportionate effect on prices, but as usual in inflationary times inflation itself affected different products at dissimilar rates of increase.

The connection between timber, iron cannon and naval supremacy had been the cause of a little-known triumph in the reign of Henry VIII. The great 'Second Bronze Age' – the age of essential bronze ordnance – lasted until the English produced iron guns that were safe for the user and much cheaper than the bronze alternative, which cost ten times more than iron guns. (This 10:1 cost-ratio was, intriguingly, the same as that between bronze and iron swords in about 1200–1000 BC.)

The English monopoly in cast iron cannon lasted only till the end of the sixteenth century, about fifty years; by 1600, ironmasters in Sweden and the Netherlands learned how to make safe iron cannon cheaply, like the English. The French, through lack of expertise, or for reasons best known to themselves, preferred brass or bronze cannon until Colbert began to price such guns against cast-iron alternatives in about 1670. The same second Iron Age never occurred outside Europe and the Neo-Europes. In Asia and parts of Africa, bronze cannon were used until the late nineteenth century. But cast-iron guns were among the great unrecognized assets of Tudor England. Their virtues compared with bronze are generally little known today, but were as vital as the difference between bronze and iron swords

There had been a near-panic and a sudden new demand for iron (and therefore timber) after Henry VIII's breach with Rome. Both cannon and shot had traditionally been imported from Liège in Burgundy and from Germany for a century before 1540. After that date, the Emperor Charles V, whose Aunt Catherine had been divorced by Henry, refused to allow any export of ordnance to his erstwhile Catholic, now Anglican, customer-uncle. So the men of Sussex developed the casting techniques for iron cannon, and they also began to cast iron cannonballs to replace the more costly earlier wrought-iron or stone alternatives.[3]

Later, in the 1770s, the dearth of charcoal and therefore of iron in Britain arose at a time when a serious land-and-sea war in America was imminent and likely to become grave. Ordinary commercial and industrial demand for iron was increasing every year because of what would come to be called the Industrial Revolution. The chief American suppliers of timber, charcoal and iron were likely to be restrained from trading with the Imperial enemy. Because of the War of Independence, United Kingdom access to an iron supply then equal to British production had to be largely forgone, while European country after European country joined the ranks of America's allies and friends. By the end of the war the Baltic trade was also denied to the British. France, Spain and Holland were at war with Britain, and there was an 'Armed Neutrality' between Russia, Austria, Prussia, Denmark and Sweden. As Britain used more iron and owned more ships than any other country in the world, the situation was, to say the least, critical.

By 4 July 1776, the day of the Declaration of Independence, it had already become obvious to the perceptive at home in Britain that there was a crisis in the production of charcoal, producing in turn a severe shortage of iron. The estranged colonies had been producing as much or more iron as the Home Country, and the Colonies had many times as much growing timber per head of population – perhaps a hunded times as much in 1776.

*

The coal solution to the shortage of timber (and therefore of charcoal) was unique to Britain. Every other nation in Europe lived and fought its wars for the next forty years, until 1815, on a largely wood standard. In addition to the early use of house-coal in place of wood, the English in the last quarter of the eighteenth century learned how to use coke in place of charcoal in the smelting of pure iron.[4]

III

When wood for fuel became short, England, almost alone in Europe, already had a domestic alternative to hand – coal. Although Roman soldiers had, it is now claimed, used surface coal to warm

their quarters in the draughtier parts of Britain, the medieval English were the first serious coal-users in Europe. The first few coal-ships had arrived in the Pool of London before the Norman Conquest; London had then and later much the largest concentration of people in England, with about 10 per cent of the national population, and, of course, it was much the richest proportion – as it is today. This remained the case throughout the Middle Ages, and coal became a better bargain than wood for burning in London, both being brought in by water. Since the accession of Queen Elizabeth in 1559, firewood had become difficult to find and expensive to buy. But although efforts to burn more coal instead of wood were inhibited by royal and noble objections to 'smog' – recorded as early as the thirteenth century – it was inevitable that coal should replace wood in the City of London. James I had enacted two regulations: first, for safety reasons, London houses should no longer be externally faced with wood, and second, bricks, tiles and glass should be baked in kilns using coal, not wood. James's regulations may have saved fuel but they did not save the City from the Great Fire.

So London became 'the Smoke' during the seventeenth century as the result of the English shortage of wood. In 1605, over 60,000 tons of coal were burnt; in 1649, the year of King Charles's execution, the tonnage had more than doubled, to over 130,000 tons; by 1700, the 'burn' had more than doubled again, to over 300,000 tons, leading to even more smog.[5]

Coal became the preferred fuel for brick, tile and glass works, for dye manufacture, salt evaporation and other industries whose origins had been encouraged by Henry VIII before wood became scarce. Some of these new industries of course originally burnt wood, not coal, but entrepreneurs were moved by the market to relocate near navigable water on which coal could be carried.

This new source of energy inspired many new enterprises, and the story of sea-salt illustrates the economic power of a high-calorific-value fuel. Before about 1550, salt was evaporated from the sea in flat, shallow pans in warm, sunny places, and England was a net importer of salt, though some sea-salt was produced in the warmer, sunnier southern counties. A century later, evaporation was so well established as an industry employing coal, and the

total amount of salt made so great and its manufacture so cheap, that a considerable export trade developed.

Alongside the new export trade in crude salt, there was an equally valuable export of fish of various grades and these two, salt and salted fish, would be connected for years before refrigeration began in the 1880s. In the later 1700s, rock salt from Cheshire became another export from Liverpool and salted fish were also widely exported as one of the few sources of protein that could sustain long voyages.[6] Salt, which for centuries had been laboriously and not always successfully evaporated in the southern British summer, or imported, became a new source of wealth when coal was used in place of unreliable British sunshine.

The manufacture of brick and glass before 1600 had been tied to supplies of wood so that clay for bricks or silica and lime for glass had to coincide geographically with timber to make possible economic manufacture, as in Roman times; coal changed the topography of these and other trades. Coal – worth double the calorific value of the best fuel-wood, and with combustion much easier to control – could now be transported to any town on or near navigable water. Contractors set up brick kilns and glass-works near their customers, not, as formerly, near a supply of wood or charcoal. Although building-quality wood was still needed for housing, much as good lumber is still needed today in brick-and-tile houses, new houses from about 1620 onwards were increasingly of brick, stone and tile rather than of wood, wattle and thatch. Later, from 1700 onwards, those new and fashionable buildings arose in every city; building sites are often depicted in contemporary pictures and these nearly always include discrete kilns making bricks or tiles for local builders nearby.[7]

Coal-fired ceramics were more evenly fired with coal and included pots, mugs, cups, jugs, plates and saucers that were of pottery, not yet of porcelain, but glazed to deliver a waterproof finish, without which no vessel could be safely used for food or drink. Although pewter for the poor and silver for the better-off were in wide use as early as the reign of Henry VII, and noted as 'normal' by foreign observers, including an astonished Venetian ambassador, practical everyday ceramics preceded true European

porcelain by two centuries. These ceramics were cheaper as well as better when coal-fired.[8]

Dyestuff manufacture was of course wholly based on animal or vegetable sources until the nineteenth century, and it was a local industry usually sited near textile manufacture. But following the introduction of coal by 1700, the coal-fired dyestuff industry was able to operate in larger units, as could brewing and lime-burning. Coal had the virtue of allowing new industries to grow naturally without always having to relocate because of wood supplies. It could be said that the use of coal was an impressive new economic factor in Britain's favour, which encouraged new local industries.

Coal was dug in outcrops, as were stone, clay and lime. Mills were built on every river with sufficient flow; alum, lead, salt and marl were exploited, often in small units. All sorts of industries were built on rivers, which provided both power and the only economic freight transport for both raw materials and finished goods. Road transport did not improve, and a town without a nearby river or canal found itself at a great economic disadvantage. Other towns with water-power, and often near a source of navigation, as in the West Country, became new centres of the wool-cloth industry which expanded from its traditional homes in East Anglia and Kent. Wholly new trades grew up, sometimes brought by religious refugees from Europe; including 'felt, thread, and lace making, silk weaving, engraving, the working of silver and the manufacture of paper, leather, needles and domestic glass'.[9] All these industries – new to England – were the product of the ferment of the Renaissance and Reformation. But not generally noted is the importance of coal in their development.

An important philosophic change followed the Dissolution of the Monasteries, which was completed in the 1540s. Educated men no longer exclusively served God and their neighbours as monks, whilst women previously in enclosed orders could now become wives and mothers. Natural abilities and acquired learning were now often devoted to Mammon, in the most economically significant way. Not since the Fall of Rome had so many pursued personal gain by peaceful means. Material distinctions grew up between England and the rest of Europe, where much economic

enterprise was still inhibited by religious or post-religious corpora-
tism, and what would become *Etatisme.* England also benefited,
because the young industries into which men went to make their
fortune were more often than not coal-fired. Trade also attracted
the adventurous, and the energy of individual Englishmen was
unleashed. Adventurers were devoted to international trade in the
Eastern Mediterranean, the Orient and Muscovy. Wool and cloth
were even sold, at some personal peril, to the pirate-rulers of the
North African Barbary Coast.

Equally, there was a development in the opposite direction.
Whilst many Englishmen found opportunities in manufacturing
as free men for the first time in history, few looked after failure,
and the incapable suffered. Because the poor became a charge on
individual parishes, the unemployed were discouraged from moving
to other towns or villages. Displaced by agrarian changes, men
often became vagabonds and beggars and starved, and Guilds used
their powers to prevent entry into the better-paid occupations,
protecting their own, as always. But as important as the social
changes was the new commercial use of thousands of acres of
woodland previously owned and managed by monasteries.

What no one now knows is the proportion of the economically
unfortunate in, say, Shakespeare's England. These people had, of
course, once been looked after by religious houses, or by their
Manors. In the seventeenth century the surplus poor would go
to the West Indies and Virginia, and those of the correct beliefs to
Ulster and New England. But before 1600 there was little opportunity for emigration. The measure of the suffering of the poor is
and was as unknown as their relative numbers.[10]

IV

By 1660, when James I's grandson, Charles II, succeeded the
temporary Republic of Cromwell and restored the Monarchy, the
English timber shortage had moved from chronic to acute. Coal
had taken over from wood in London and in other cities close to
water transport, but it was not yet favoured for baking high-class
bread or drying malt for beer of good quality. These processes were

more successful when near-smokeless charcoal was used, sulphurous coal affecting the taste of most foods. Not until the widespread manufacture of 'clean' coke from suitable coal was there available any fuel pure enough to use in the production of good food and drink. Making bread and beer of the highest quality was too important to be entrusted to any fuel except charcoal, which was preferred over any other source of heat – even coal-gas – in really high-class kitchens in some parts of Europe as late as the luxurious times before 1914.

In England, the only practical alternative to wood was coal, and the earliest coal used lay above ground or in shallow pits; such easily mined coal was soon exhausted, however. Most of the coal in England was some way underground, and the geology of English coal measures nearly always involved water, and water implied pumping. No significant tonnage of coal could therefore be dug without some means of powered pumping.

Before steam, there were alternatives, but none was powerful and none of much use for deep pumping. Early woollen mills were driven by human, animal or water power; there were treadmills for slaves or criminals in ancient times, pumping water, crushing ore, grinding corn. After AD 1200, there were windmills in flat country in Europe in those areas without flowing water, and watermills on many streams and rivers had been noted in the Domesday Book in 1087. But wind- and water-powered industries, naturally widely dispersed, were difficult to concentrate and were expensive in capital cost per horsepower (hp) generated.

The building of early steam prime movers was therefore originally made essential in England by the shortage of wood and the consequent need for deep coal and therefore pumps. The earliest pumps were for draining mines, and these were installed from before 1700, the chief partner-inventors being Savery (1695) and Newcomen (1698).[11] These early pumps were 'atmospheric' and there was no effort to conserve heat, the whole, huge cylinder being cooled after each 'stroke' by a jet of cold water. As a result, both efficiencies and outputs were very low, the efficiency before 1720 being less than 0.5 per cent and the output about 1 hp. Steam at only atmospheric pressure raised the piston to the top of its stroke, largely by counter-weight. Condensation was by cold water.

The resulting vacuum allowed atmospheric pressure to force the piston to drop to the bottom of the stroke, raising the water.

The growth of steam-powered pumps in the seventy years before James Watt is almost unknown to non-specialists. In the 1720s Newcomen exported machines for use in France, the Rhineland, Belgium and Saxony, and installed them in many places in Britain; perhaps a hundred, half abroad, half at home, were in place by the time of his death in 1729. Ironically, all the iron used in making these steam pumps was derived from forests, and so obviously was the wood used in their construction. They had a high wood content for an efficiency of only 1 per cent.

The limitation of the original design was almost crippling, even when manufacturing was refined and improved by John Smeaton. He had far greater production expertise, and was no longer dependent, as Newcomen had been, on carpenters, blacksmiths, wheelwrights, and even saddlers, who made the early leather seals. Smeaton used boring machines whose bores were far more accurate than was possible with the crudely wrought iron cylinders of the early Savery–Newcomen pumps. But however accurately machined the parts of a Smeaton pump, efficiencies could never be greater than 1–2 per cent because of the cooling of the cylinder at every stroke. None the less these crude pumps made possible the first deep-mined coal and increased the quantity lifted from virtually none in 1690 to more than 4 million tons in 1776. To this must be added perhaps 1 million tons from surface mines each year.[12]

Coal saved wood in a dramatic way. In 1690, less than 1½ million tons of coal from surface or undrained shallow mines were burnt annually in Britain and the use of coal saved about 40,000 acres of mature woodland. On a 100-year cycle, this would mean 4 million acres; on a 125-year cycle, 5 million acres. Even if non-native quick-growing softwood varieties had been planted for fuel-wood, about 1 million acres of woodland would have been needed to replace each million tons of coal burnt each year, on a self-sustaining basis. In 1776, with 3 million tons of coal being burnt, between 10 and 15 million acres of woodland were saved.

The weakness of the Smeaton pumps, even when superbly constructed, was philosophical, not mechanical. There was a single-acting piston in a cylinder which lost all its heat at every stroke. The loss of heat was of great importance, yet apparently unrecognized, although it breached known Newtonian laws about the conservation of energy, and it needed an abstract thinker as well as a craftsman to solve the weakness. The first man to do so was James Watt, an instrument maker employed by Glasgow University who was already halfway to being a natural philosopher as well as a craftsman. He solved the problem of the loss of heat by using an external condenser.

After various – inevitably barren – negotiations on funding, Watt worked in partnership with Mathew Boulton of Birmingham, and the firm Boulton & Watt (along with the master patent) was in place from 1775 to 1800. Ultimately, the machines were double-acting, the motion rotational with the speed controlled by the now-familiar butterfly governor. Almost anything could be driven by stationary steam-engines: water-pumps, air compressors for blast furnaces, winding gear in mines, entire cotton mills, with elaborate and – to our eyes – dangerous unguarded belting to transmit power locally. The efficiency of the Watts machines made a great impact on a nation needing more power. As horizontal boring machines became more effective, and steam more efficiently used, Boulton and Watt engines were ultimately three times as kinetically efficient as Smeaton's best, and nine times as efficient as Newcomen's best steam pump. If he had used higher pressures ('dangerous', said Watt) efficiencies would have risen further.[13]

By 1800, the United Kingdom of England, Wales and Scotland had a population more than twice that of the United States and half that of Revolutionary France. France was rich and the United States very rich in woodlands, and both economies were still driven by wood, wind and water, like those of every other nation in the world with the exception of the United Kingdom, which was partially dependent on coal. (Coal-deficient Ireland joined the Union in 1801.) In 1800, 6 million tons of coal were dug, distributed and burnt, equal to 120,000 acres of timber clear-felled every year, or a rotational area of at least 24 million acres. This area did not, of course, exist, since the United Kingdom produced all its own wheat, beer and beef and every other temperate food common at the time, and the margin was tight. Both wind- and water-power had reached their (then) very low upper limit, equivalent to a few thousand horsepower in England.[14]

Sugar, tea, coffee, rice, indigo and raw cotton and other traded goods were almost wholly transported in 8,000 British merchant ships, guarded and often convoyed in wartime by the Royal Navy. These operations had a high wood and wood-fuel content and an unavoidably warlike nature, while the British war effort also included a huge output of ordnance as well as of small arms, and gunpowder to match. This war effort in 1810 has been quantified at about 20 per cent of the output of British industrial effort, and together with sea transport and trade consumed nearly 40 per cent of the gross tonnage of available fuel. But Britain at this time was like the United States in the 1940s, capable of rapid industrial expansion, and there was no severe inflation before 1808. It was also of course true that any product with coal in its manufacture would be cheaper in Britain – and only in Britain – than if made using wood.

The United Kingdom used more coal much earlier than any other country in the world – it was more than a century earlier, compared with most other economies. The UK use of coal in 1800 amounted to many times the use per head of any other country, as much as 90 per cent of all the coal burnt world-wide; more than half was used in industry, not in the home. But the use of the wood-wealth of the American colonies alone gave England the opportunity to trade all over the world, her merchant ships protected where necessary by a generally effective Royal Navy. Without coal and North American timber, the British could never have supported the new industrial manufacturing base and a rising standard of life as well as a rising population when timber ran short. Nor would there have been an increase in real national income per head greater than that in any other European economy during the eighteenth century, nor of course victory over Napoleon, nor the second British Empire, nor the splendours and miseries of the nineteenth century, with its wholly characteristic carboniferous capitalism.[15]

V

If there was one over-arching reason why the British-American Colonies were so materially prosperous, in contrast to the Home Country, and if there was a single explanation for the early wealth of the Colonies, it must be timber. With no commercially useful timber on the north-east coast of British America, history would have been very different throughout the Colonies. If there had been no indigenous species of soft or hard wood as good as or better than European equivalents, fishing, whaling, shipping, ship-building, exports of masts, timber, tar, iron and the lumber trade would have been unimaginable in the form that they took. British-Americans would never have been as rich as they were, often wealthier than the relatives they left behind.

Settled in a timberless land, colonists would have been as dependent on the ambiguous goodwill and dubious altruism of the Home Government as were other colonists of other powers in the New World. The only substantive export to Europe would have been Government-controlled tobacco, a trade only made profitable because it had been made illegal to grow commercial tobacco in England. Most indigenous whites would have been poor, as poor as poor whites were in Virginia when their land became exhausted by growing too much tobacco for too many decades.[16]

When they crossed the Appalachians, Americans found a wealth of timber. Most of the land east of the Mississippi–Missouri Valley was forested, and most of those who moved west before the Civil War settled in wooded areas. The virgin forest provided the settler-family with all its needs. A sort of sheltering tent, made of timber fronds, could be built in a few hours; a more permanent log-cabin could be constructed in a few days with ample standing timber all around.

On both sides of the Appalachian chain, log-cabin walls would be of whole or split logs, the floor and door of crude planks, the fireplace and chimney of clay if that were available, or later bought-in bricks to safeguard against fire. Clay or bricks had often to be hauled many miles.

The green wood of new log-cabins required fires burning for a long time, except in high summer, to dry out what had so hastily been put together without any seasoning. The gaps between the dried-out logs allowed in wind and cold, so they were filled with clay; later – often much later – walls were lined with properly seasoned, sawn and planed board. Fences were also made of timber, a good man being able to cut and split 200 rails a day from felled logs or in the same time cut and shape 100 stakes. Much later, because he was a quicker worker than average, the young Abe Lincoln was contemptuously known to citified Easterners as 'the Rail Splitter'. Counter-intuitively, this would be to his political advantage.

Tree-trunks could be used to make bridges over brooks and streams, paths or 'board roads' across marshy ground, even drain-pipes when hollowed. Because the virgin forest formed no perma-nent part of the plans of most pioneers, domestic animals were encouraged to forage in the woods, once the quarter-section had been fenced. Pigs especially benefited, almost keeping themselves; some went wild and reverted in appearance to the wild boars of Europe, whence ultimately came their genes. Fencing was at first glance quite a task, since the quarter-section of 160 acres allocated to pioneer settlers needed at least 2 miles of fencing, the minimum being an 880-yard square. Two miles of fencing, in turn, would have needed nearly 1,200 stakes and twice as many rails; presum-ably this load could have been halved by sharing the burden with adjacent neighbours. But every State had different local rules about livestock-fencing and local laws, different in a number of ways from the Common Law brought over from England. A unique American contribution was the picket-fence, a free-standing affair, about 3 yards long, which would render any area relatively stockproof except against horses, which could normally jump them, as was proved when picket-fences were used in an attempt to protect infantry against cavalry charges in the Civil War. Fortu-nately for the casualties of that brutal-enough conflict, barbed wire came later. But, in the 'peaceful' post-war West, horses learned to jump both wooden fences and the newly invented barbed wire, and had to be hobbled.[17]

Virgin deciduous forest was often responsible for building fertility in the soil, a natural wealth banked before any white men arrived, because of the annual leaf-drop. Some species, like beech, were more valuable for this purpose than others. Today, foresters know which are the valuable and the less valuable species for enriching the soil. This should have been a factor in valuing forest-land that settlers intended to turn into arable, but there is no literature on the subject.

More immediately important were the proximity of wood and the nature of the market for wood. Because there was so much of it about, felled timber was not regarded as having a greater value than as firewood, and most wood was burnt if not otherwise used. A cord of wood (8 × 4 × 4 feet, or about a ton), was the unit for fuel, and 40–60 cords per acre the norm. Proximity to a town produced an immediate local market for firewood but a nearby railroad or river would mean that locomotives or steamboats would be the consumers. There were more valuable uses, however. Some bark went to tanneries; some large trees could be planked for lumber for building purposes, always provided the transport was available; white oak was needed for barrel-staves in which every-thing, from meat to tobacco to flour to pork, was packed in those days.[18]

A more valuable – and more ecological – solution was to establish a sawmill nearby, to add value to what would otherwise be firewood and therefore sold cheaply. But a sawmill would need good transport, several months to set up, capital and, above all, a supply of water-power – or sometimes, if more capital was available, a steam-engine. Despite these difficulties, sawmills were established, timber was seasoned, and many communities benefited. A man with a sense of husbandry might prove a role model, providing a degree of leadership and conducting himself in a way that was kinder to nature and to his neighbours than was customary.

In the absence of any market for wood as lumber or fuel, trees would be burnt in huge heaps, the resulting potash-rich ash being more valuable than the standing timber from which it was derived. Not much is heard of charcoal-burning in pioneer Frontier areas, nor of its main outlet, iron smelting; this presumably occurred at a later, more sophisticated stage than that of the pioneer-settler.

VI

North-East America was the first place in documented history where ordinary competent Europeans had no need to go hungry. In other words, every European in North-Eastern America, however poor, had plenty to eat for the first time. This was not immediately apparent, but even in New England, with its thin, stony soil, the Pilgrims were eating better within a few years of their arrival than they had ever done at home.[19]

The Amerindians the Pilgrims met had no tradition of animal husbandry, for the best of reasons: there were no domesticated animals. If not hunter-gatherers, Amerindians in the North-East were simple growers of plants who used maize as the staple starchy food that all humans need, whether in the form of American maize (corn), European wheat or Oriental rice. Maize was, of course, the only American plant available to produce a staple alternative to cereals.[20] There was probably little man-cleared land, just a few natural groves in the forest, or along a river where floods or beavers building their dams might have cleared flat alluvial patches. It is easy to grow maize without any kind of cultivation; a small hole in the ground made with a stick for each maize kernel and a watch for pests suffice. Before the white man, there were relatively few animals to trample down the growing maize, fewer to eat the green plant, and none to prevent its spread through much of the Americas, from the St Lawrence to the Plate.[21] In places in pre-Columbian New England, there were just a few moose and other deer; turkeys on the ground, trout in the rivers and lakes. This was an epic sylvan scene, pastoral in the best sense, and the small Amerindian population pressed not at all harshly on the environment which was home to species now extinct.

Of all the changes wrought by white settlers in New England, the destruction of forest habitat was the most damaging to the Indians. Building dams, millponds and races for timber extraction interfered with the natural flow of streams and rivers, and killed fish and threatened other species. River-engineering did as much damage as clearing forests to provide the white settlers with land. The Pilgrims managed the flow of streams and rivers, and their 'improvements' destroyed the habitat of fish, birds and mammals and also diminished the food supply of the Amerindians. The settlers, once tolerated by the natives, came to be hated. Possibly more detested than the settlers themselves were their methods of fishing, trapping and farming and their dependent and domesticated European animals – cattle, horses, sheep, goats and pigs – which were unknown to the Indians. As white population pressure mounted against the natives, they regretted the kind welcome they had extended and warned the Europeans not to encourage yet more immigration, nor venture deeper into Indian lands. The Europeans (mostly British) ignored these requests and Indian wars, at first very small affairs, mere brawls, ultimately became commonplace. The original cooperative relationships ended and the whites pursued a policy of divide and rule.

Although this is the politically correct version of early New England history, it is selective. In fact, most 'Indian' wars involved other white men, sometimes of the same nationality, sometimes the European enemies of the British, first the Dutch, then the French. Nor was slaughter by aggressive whites largely responsible for the mortality rate among Native Americans. Ninety per cent of all Indian deaths in the century after the *Mayflower* can almost certainly be charged not to warfare, but to the diseases that the whites brought with them. Smallpox, tuberculosis, measles and malaria were the great killers – unknown in the Americas before the arrival of the Europeans. Smallpox had first occurred from occasional contact with European cod fishermen and was already widely dispersed among Indians between the St Lawrence and the Hudson and east towards the sea, before the Pilgrim Fathers arrived in 1619. Where smallpox led, other diseases followed, until most Indians were wiped out; those that survived developed immunity, or the diseases themselves attenuated.[22]

Beyond the disease factor, it is also true to say that Stone Age Indians found their lifestyle impossible after only a few years adjacent to white settlements; and the great forests that had once supported a wonderful bio-diversity of life became as relatively barren as they are today. Turkeys, present in huge numbers before 1600, did not long survive the invasion from Europe, and it was dogs and gunshot that diminished their numbers, unless the white men inadvertently imported diseases fatal to turkeys in their European chickens.[23]

British North America differed markedly from other European Colonies further South. First, there was no organized, splendid if savage, Amerindian culture, as existed in Mexico or Peru, which had perforce, in the philosophy of the time, to be destroyed by the Conquistadores. Indigenous natives in most of the British Colonies were of the Stone Age. They were primitive hunter-gatherers or subsistence growers of maize with no knowledge of animal husbandry. They had no draught animals to reduce the burden of drudgery. Colonial Amerindians were also thin on the ground, thinner still after being exposed to the diseases of the white men.

Second, British North America offered whites almost no opportunity to get rich quickly. There was neither gold nor silver and every export – fish, furs, timber, and so on – had to be farmed, grown or hunted. The introduction of European flora and fauna was equally labour-intensive. The consequent struggle for existence tempered the character of the white population in the long run.

Third, this was a place where Europeans multiplied extensively. The net increase in population meant a million whites in 1720, more than 2 million by 1770, and nearly 4 million by 1800, the last figure signifying an increase of nearly 300 per cent in fifty years. These statistics should be compared with a net gain of 80 per cent in the United Kingdom between 1750 and 1800, the highest in Europe, admittedly, but moderate compared with North American demographic increases.

A great increase in numbers was reinforced by a great increase in physique. British-Americans became larger, fitter and stronger than their European-British ancestors and collaterals, and survival amongst babies and young children exceeded European rates by a wide margin (a major factor in the large demographic gain). This survival rate was partly because American English, Welsh, Scots and Scots-Irish (more than 90 per cent of white immigrants before 1800) were not packed into large, crowded towns like London, where disease killed more than half of all infants before their third birthday. Since British-Americans were better fed, pregnant and nursing mothers and babies had better resistance to perinatal and childish ailments. All this led to a feeling of optimism, or did a feeling of optimism lead to the high birth-rate?

*

The region between the Appalachians and the Great Valley, now occupied by all or part of fifteen States, contained growing timber covering an area larger than the whole of France, Italy and Switzerland combined. This timber, if burnt, would have had a thermal capital value of 5 billion tons of coal equivalent, or a sustainable value (on a 100-year cycle) equal to more than 50 million tons of coal a year. In fact, the timber was used for houses, farmsteads, railroads, ships, bridges, wagons and furniture as well as for fuel, and some was simply destroyed to make agricultural land. The United States was launched into the Industrial Age uniquely with wood, not coal.

There are four contributions to mass-production (and increased wealth) that are now, 200 years later, identified as obvious, but their importance was resisted by those without the intelligence or experience to benefit directly from their adoption. Yankees assessed their value by trial and error, not by pure science; nor were their virtues widely voiced, being only first taught formally at MIT in the 1870s. The four are:

- the adaptation of power to machinery;
- the principle of interchangeable parts;
- machine-driven tools to make components; and
- original (appropriate) design so that unskilled labour can be used in assembly.

In every one of these developments in the Colonies (later the States), New England was a leader and ample wood played a major role.

Water-power was employed in New England in the lumber industry as early as 1628 and water-power (thanks to wood) energized many industries for more than 200 years. With the advent of steam, cheap wood fuelled American steam for far longer than anywhere else.

Machine-driven tools were adopted more enthusiastically in New England than in Old England and often made cheaply and quickly out of wood, like clocks of the same date.

Interchangeable parts were developed as a principle by Eli Whitney prior to 1800, and before others in Europe: Wilkinson – textiles; Colt – revolver parts; Root – ironworks; Blanchard – copying-lathes, and others.

The use of *unskilled labour* was forced on American manufacturers because of a shortage of *skilled* labour. The skills of machine-minding were unknown and had to be learned. Historically, skills of every new technique have to be learned, and in the process, existing proponents of the old technology become automatically 'de-skilled'.

'De-skilling', which occurs and re-occurs throughout history, is often sad for the individual, but new skills were resisted less in New England than in the Mother Country, and the tradition of adaptability became one of the strengths of the United States. In parallel was an absence of frame-breaking Luddites, whilst throughout the Old World there had been powerful craft-resistance to new methods, and it was to the benefit of the new countries that there were few to resist factory novelties. However painful in the short term, even over a generation, the adoption of new processes ultimately favours workers far more than does resistance. The difficulty is to know which new technology is inevitable; only the market can tell, and sometimes only with hindsight. So only a few may benefit immediately, and it may take years for new efficiencies to profit economies and longer still to 'trickle down'.[24]

The chief problem facing the Colonies and the young United States was not lack of resources, nor even a lack of people, but a need for good transport for the conquest of distance. The country was enormous, even alarmingly so for people who travelled on land at the speed of a horse. There was nearly as much land in the thirteen original States as in France and Italy combined, but the new country was not compact, being than 1,200 miles long and a few hundred miles wide at the most, more the shape of the leg of Italy than of France. It is a mistake, however, to believe that the need for transport systems was clearly recognized at the time. The Spanish-American Empire was challenged by the same problem of internal distance, and it was solved by neither the first European settlers nor their successors. The United States was lucky enough to be young at a time when steam, canals and railroads were about to become viable, and North Americans took advantage of every technological improvement on offer.

Before Independence, nearly all long-distance transport was by ship and boat, powered by sail or oar; but there are pleasant historic connections about the early development of steamboats. There was a partnership between Robert Livingston, who had signed the Declaration of Independence, his brother-in-law John Stevens, who was an engineer, and Nicholas R. Roosevelt, great-uncle of the future President, Theodore. These three acquired a monopoly to navigate the Delaware River, but their steamboat never worked economically. Then Livingston became American Minister in Paris, and there he met Robert Fulton who was trying to sell Napoleon the idea of submarines and torpedoes to sink the Royal Navy.[25] These missiles only existed in Fulton's vivid imagination, however, and even before Trafalgar, Napoleon had gone off to conquer Central Europe. Fulton returned to the United States in 1807, and backed by the venturesome trio of capitalists, he built the first steamboat on the Hudson, a paddle side-wheeler which made the 150-mile journey to Albany in thirty-two hours. But the Fulton river-boat was badly designed, and even when rebuilt did not last long and never made money. Monopoly was called in to save the commercial situation, and there followed attempted monopolies on many rivers, including the Mississippi, before they were ruled unlawful by Chief Justice Marshall and the Supreme Court in 1824. Before this vital ruling, the first technically successful (non-Fulton) steamboats had been built on the Ohio, with a shallow draught, high-pressure boilers and engines mounted on deck. By 1830, there were over 200 such steamers on the Western rivers, later to become the travelling palaces of legend, complete with professional gamblers, loose morals and steam organs. These river-steamers were 90 per cent wood and they burnt their own weight in timber about every 1,000 miles. By 1860, there would be ten times as many wood-fuelled steam vessels on the same rivers – 2,000 of them – and wood, not coal, was to be found at almost every point along their course.

In the same order as in England, but later, canals and railroads became the one answer to the American internal transport problem after 1790. No one today can imagine the difficulties involved in moving heavy inland freight far from rivers before there were canals or railways. It was largely a freight problem, because anyone in a hurry could ride a horse. After the arrival of canals and railways the use of roads for freight was not viable until the perfection of the internal combustion engine. Every experience since the great Persian roads built before 500 BC had confirmed the dubious nature of any claim to economic benefit for roads for freight, in contrast to military or political advantages. Fine for runners or horse-riders, and for moving soldiers, ancient roads, even the superb Roman roads, were of little use for freight wagons. Because Roman harness was so inept and wagons so inefficient, and because draught animals had to be fed, the cost of a wagon-load of grain doubled every 100 miles. After about AD 1200, European harness and wagons had improved, but roads – very expensive in upkeep – had declined in quality and ultimately some had to be supported by tolls. American draught animals and vehicles did not suffer from any harness-imposed limits, but there were similar problems to those in Europe. In the United States, horse-drawn freight transported by road cost ten times as much as horse-drawn freight transported by water in 1800 – forty times the cost of steamboat-transported freight in 1830. Before railroads, canals became of greater importance than roads and some of them evaded or defeated the Appalachian barrier. But many of those supported by individual States suffered from financially unsecured paper and later could not compete financially with railroads, and investors paid a high price for the eight States that defaulted.

Other canals had uses other than transport; one was the La Salle, which allowed the rapidly growing city of Chicago to dump its sewage into the Mississippi River, via the Illinois River, thereby keeping its own lake shore clean for fish. The nominally more fragrant reason for the La Salle Canal, going south, was to link the Lakes to New Orleans, whilst the Erie Canal, going east, joined the Lakes to New York.

The Erie was the most commercially successful of the canals, allowing lakeside States to deliver grain to New York via the canal and the Hudson River. To the surprise of those without any historical sense, it was revealed that grain from lakeside Ohio was cheaper delivered by canal to Charleston, South Carolina (more than 1,200 miles) than by road from inland South Carolina (100 miles). One effect was to inhibit Southern wheat culture.

The Erie Canal repeated the experience of the Emperor Nero (d. AD 58) and his canal parallel to the Tiber, or that of the medieval Venetians, who imported grain from Cracow by water – over 5,000 miles – instead of less than 250 miles by road via Austria-Hungary.

Before the building of the Erie Canal (1825), New Orleans was the natural outlet for the huge basin drained by the Western rivers. By 1830, New Orleans found itself losing much important bulk trade, largely grain and lumber from the then North-West. It was not until the river States were developed that the Mississippi ('Ole Man River') regained economic importance. There was also a political price to pay before 1860. Because of the Erie Canal, commercial interests in the old North-West tended to align themselves with the East in slavery debates and this was in the end a higher cost than any earnings lost from freight not passing through New Orleans.

Even after the arrival of railroads, river and coastal steamers were still vital in the East, not only on the Hudson but from New York north to Boston, south to Chesapeake Bay, and water transport competed vigorously as far as cost was concerned. Timber-fuelled steam-power was also essential for the economic movement of bulk cargoes on canals. Most American canal barges were much wider than horse-drawn narrow-boats in England. They were of wood, with tugs fuelled by wood, as were the river steamers.[26]

All inland water transport has by definition to be relatively slow, even compared with slow early rail movement, and inland waterway freight has to be largely free of the major constraint (or tyranny) of time. Given an absence of time constraints, canal and river transport was still far cheaper than rail in 1860. But no one should imagine that water transport was run by cost-reducing philanthropists. Before railroads became an important deflator of costs, both river and canal rates were ten times what they were twenty years later. It was steam that reduced both railroad and inland water costs, and railroads were half the British cost because of cheap American timber for construction as well as for fuel.

Lower internal transport costs in the huge new country helped double the economic growth rate of the young United States. Between 1800 and 1840, economic growth had been running at a rate not very different from England's. After 1840, it increased so much that American growth rates were nearly double those of England in 1860. Recovering from Reconstruction, the US growth rate was three times that of the United Kingdom in 1880. This was largely due to the reduction in transport costs following the triumph of wood-fuelled steam, which was of greater importance in a huge country like the United States than in the smaller United Kingdom. The American West that Thomas Jefferson (d. 1826) had thought would need a millennium to fill with a population big enough to exploit its bounty, was almost wholly settled before the centenary of his death.

*

The same sort of progress in productivity was made at sea. In 1860, American export freight rates were one-twelfth the rates for the same goods in 1800; this reduction was due to improvements in the size and efficiency of wooden ships, still driven by wind, steam having a more immediate effect on inland waterways than on the oceans. By 1860, freight by steamers on rivers or canals was just over 1 per cent of the cost of freight by wagon in 1800. River rates came down by four times, if a raft floating down to New Orleans in 1800 is compared with a steamer doing the same trip in 1830. The raft was broken up, never to be returned; the steamer came upstream at less than half the speed it travelled downstream. River-freight productivity doubled again between 1830 and 1860, whilst railroad productivity increased by even more after the Civil War. But even when railroads were fully developed, they could not compete with a bulk cargo, free from the demands of time, travelling on the great rivers flowing to the Gulf. This is still true of river-freight today.

The first American railroads, like those in Europe, were to be found in quarries and mines; as in Europe, wagons were first hauled

by horses, mules or cables. Other early railroads were powered by
sails mounted on trucks; some even used a horse on a treadmill,
as was tried and (predictably) failed at the Rainhill Trials in 1830
in England. American railroads began more quietly than in England and the railroad was usually conceived as a freight line. Then
it was found that advantages in speed compared with water-routes
made passengers important. Turnpike roads and competing canals
were not normally allowed to hold up railroad development, as in
the Old World, and there was usually enough empty land not
to necessitate the expensive financial compensation which made
railways so costly to build in England.

'Free' land played an important part in American railroad
building. Twenty million acres (31,520-plus square miles), an area
larger than either Scotland or South Carolina, was 'given' to
railroad promoters before 1860. This was to encourage rapid
build-up of traffic, the 'free' land being settled by farmers and
others. In exchange, there was the right of Governments, State or
Federal, to enjoy preferential freight rates.

Three features of nineteenth-century American railroads went
beyond European experience. First, there was the force of competition, often remorseless and usually without scruple. Second,
there were marvellously inventive financial tricks, including much
'watering' of stock. While these features became more general
after the Civil War, they were by no means rare before 1860.
Third, there were no real early interconnections by rail; in 1860,
for example, rails from New York to Buffalo were in six systems,
in three different gauges; passengers were required to change carriages, while freight had to be trans-shipped. Neither a common
national gauge nor a zonal time was established until the 1880s.[27]

Nor did American trains find it odd to travel along highways;
this usually happened in towns, with the locomotive's warning bell
clanging and the whole procession advancing at the pace of a horse.
Abraham Lincoln proceeded thus, in (imaginary or real?) peril,
through the streets of disaffected Baltimore, on the way to his First
Inaugural in Washington, DC, in March 1860. Even today, trains
can be found passing through the middle of Main Streets in Middle
America, while the fabled daily Twentieth Century Limited, sched-
uled to run the 960 miles between New York and Chicago at an

average speed of 60 mph, crawled through the streets of Syracuse, as recently as 1936, at a legal limit of 15 mph.

American railroads were far more wood-reliant than the railways of any other country, even those of Russia. Early American rails were often made of wood, shod only by a thin strip of iron; ties (sleepers) were of wood, of course, as were even the larger bridges, the longer viaducts and many drainage and water-supply pipes. Wood was the fuel of almost every US railroad locomotive until after the Civil War, and by then the United States had twice the rail mileage of the United Kingdom, much smaller, of course, but hitherto the most developed railroad country. The United States probably burned ten times as much wood as the United Kingdom burnt coal, but green softwood of course has less than a third of the thermal value of coal.

The Civil War, the first railroad war in history, showed engineers what could be done with timber under the stress of battle. Near Fredericksburg in Virginia, a wooden viaduct replaced a structure burnt by retreating Confederates; it was built over the Potomac Creek in nine days. The wood for the viaduct came from local standing timber, amounting to over 200,000 cubic feet, or the product of at least 500 acres. How green the timber was, and how warped the bridge – made of unseasoned wood – became, has not been recorded.

After the Civil War, in places west of the Mississippi–Missouri basin and east of the Rockies, timber had to be hauled so far that iron was sometimes preferred for bridges, and masonry for viaducts. But there were disadvantages: wrought iron was never as flexible as wood, and masonry was often much more expensive. On the other hand, neither metal nor masonry was easy to burn. In the wilder parts of the Frontier, wooden structures were sometimes destroyed by rivals or criminals or just dismantled by the cold or hungry. Lastly, the age-old rule applied: iron smelted with wood, even in 1860, often cost more in timber than wooden material of the same strength as iron. Over time, of course, iron would be cheaper, because it would probably last longer. But before and after 1860, most iron in the United States was smelted with wood, a condition no longer true in England. During hungry winters in the 1880s, settlers even took to stripping wood for fuel from bridges and viaducts, an enterprise not esteemed by the railroads.

American railroad coaches were derived from lavish, comfortable, long canal-barges for passengers. They were too long to be manageable or safe on a conventional 'fixed chassis' when running on the rough-and-ready, light, cheap and cheerful track of the 1840s and 1850s, so four-wheel bogies were evolved to prevent the derailments that were inevitable with a long, rigid wheelbase. British railway coaches derived from stage-coaches and were usually built with compartments with six, eight or twelve seats, according to class, and with four or six wheels on a rigid chassis running smoothly on high-class permanent way. The American open coach, which was replicated in railroad cars after 1870, encouraged strangers to talk, and was inclusive of people; the much smaller British compartments were associated with much more formal behaviour.[28]

Nearly every railway coach before 1860 contained about 90 per cent wood, which in Britain was nearly always imported timber. Whilst there was little comfort in British trains, American railroads provided WCs a generation before any European country other than Russia. Both Russian and American railways, because of long distances and slow speed, were also the earliest providers of diners and sleeping-cars. There were few gangways in British coaches before the 1880s, and gangways with access to all carriages and dining-cars were not generally found, even on the best trains, until the 1900s. Fire risks were ironically increased by improvements in lighting; the progression was from candle-lamps to whale-oil lamps to vegetable-oil lamps to kerosene lamps to piped gas. Piped gas provided fuel for the worst sort of fire in accidents: the only safe lighting – by electricity – was not installed, even in new rolling-stock, much before the twentieth century.

Freight vehicles answered the trading needs of the countries involved. America soon developed 40-ton freight-cars with double bogies; Britain relied on 10-ton four-wheelers; America had automatic couplings by 1860; Britain was still using screw-couplings and buffers – designed in the 1820s – more than a century later. At least three types of shock-free wheels were made of wood and patented on both sides of the Atlantic. Locomotive fuel was originally coke in Britain, wood in America; coke gave way to much cheaper coal when the brick arch and deflector plate were invented in the late 1850s in England. As with iron-casting, anthracite from Pennsylvania was the first American substitute for wood in locomotives, anthracite being first successfully burnt in very wide grates on the Reading Railroad in Pennsylvania. But this alternative was not generally available except in the East.

The substitution of steel for iron rails probably ultimately reduced the amount of wood used in American railroads as much as any other factor. This was because iron was largely smelted with wood in the United States, whilst steel produced directly by the Bessemer or Siemens-Martin processes required a fuel with a far higher calorific value – coking coal. No steel rails were rolled in the United States before the 1860s but, when they were first used, they outlasted iron by seven to ten times. So did the rims of vehicle wheels, called 'tires', though they were not detachable, and by 1880, in most countries, new wheels and rails were universally of steel, not iron. The use of steel conveniently coincided with the US railway construction boom from 1870 to 1900. During this thirty-year period track mileage increased four and a half times, to nearly 200,000 miles, at approximately $35,000 per mile, or about 40 per cent of British cost. (At this time £1 was worth $5.) What was true of the majority of railroads was that fixed costs were a higher proportion of total costs than in most other industries, except, much later, nuclear power plants.

By 1860, fixed investment in US railroad had reached over $1 billion, and ten times this figure by 1914, when repairs, renewals, depreciation, and interest on capital amounted to about $2 billion a year, the highest standing charge in any industry. These high fixed costs (much the highest cost element in a railroad) encouraged two tendencies. First, there was apparently not much loss in running half-empty trains. Second, there were few opportunities for saving money on fixed costs so any savings had to be made on consumables, equipment or labour, which were always a smaller proportion of the total. As a result, labour tenure became entrenched in time, almost as an antidote to the low wages which purer economics would have dictated. Where there was true competition between railroads, as in much of the United States, labour-saving devices proliferated, and ton-mile and passenger-mile costs came to be lower there than elsewhere, often by a wide margin.

In many European countries, railways ran at a loss, or were subsidized as a public service, or were cross-subsidized, as in France, Germany and Austria. In the most extreme case, no one knew which part of the service was run at what profit or loss, since (often contrived) obscurity shrouded financial facts even from senior management. In the end, many railways were unable to meet the later competition of the internal combustion motor in road vehicles and had to be protected by legislation and taxation. Mature railways were run, it often subsequently appeared, for the benefit of employees, not customers. Claims made by trade unions about railway working conditions were not always well founded. For example, for over 150 years there has been a claimed connection between safety and manning on railways. Yet American railroads employed one-half the workers per mile compared with the British in 1840, one-quarter in 1860. There were several times as many people killed per mile of track on British railways as on American railroads. This was partly because of a more intensive use of track in Britain, but there have to be other reasons for such a huge difference, perhaps including the quality of both workers and management. Similar anomalies occur in other pairs of statistics.

*

One of the chief passenger sea routes in the world in the nineteenth century was the Transatlantic, as is still the case but today by jet. Now the vast majority of people travel by air, for business and pleasure; equally, nearly all return. Between 1845 and 1860, the one-way traffic to the United States by sea was never less than 100,000 passengers a year, and immigrant numbers rose to an antebellum peak of 400,000 in 1852. In proportion to the then population, immigration was equivalent to nearly 4 million a year today. The disturbance can be visualized, and the aggrieved media reaction imagined; but in actual fact, there was not much doubt about the benefits of immigration in the 1850s. This was partly because of the very low density of the US population, but largely because of the Frontier.

Immigrants went West, or Americans went West and immigrants remained in the East. If Americans went West, their place in the factories and cities of the East could be taken by immigrants. If the immigrants went West, trade would increase and the new railroads would benefit. In the years either side of the Civil War there was plenty of land at $1.25 an acre or less, the standard allocation for a pioneer being 160 acres, or a quarter of a square mile, the area called a 'section'.

It was held in 1830, before the railroads, that there could never be much White American settlement west of the Mississippi–Missouri, because the Great Plains were so seriously short of wood. By 1860, 3 million people had settled in the five wooded States, Tennessee, Kentucky, Ohio, Indiana and Illinois; in wood-deficient Kansas and Nebraska, fewer than 8,000 had claimed and occupied their quarter-sections, but some serious settlement had been made in the wooded river valleys on the Kansas–Nebraska border. What transformed settlement in the States between the Mississippi–Missouri Valley and the Rocky Mountains was the arrival of railroads – the branch network especially – after the Civil War. Before 1860, a settler in what is now Western Kansas or Eastern Colorado lacked timber for housing and fencing, even wood for fuel. The traditional wooden log-cabins gave way to cabins, like Irish-style cabins, built of blocks of sod (instead of peat).[29]

The cost of fencing was a major problem. Although barbed wire became available in the 1870s, it cost between $100 and $200 a mile, depending on geography and transport. Stakes were still needed, and they might have to be hauled more than 500 miles and cost as much as 25 cents each. So the 1,200 stakes needed for a quarter-section would cost $800 in real money, without any intermediate fencing to divide the farm into fields. Because of its high cost, fencing was not general in the West, and cattle became 'hefted' to their spread and branded with the ranch's device, Diamond T, Lazy S or whatever. Horses could be tethered or hobbled. Water was as much or more of a problem in most of the Great Plains, and shortages of timber and water were not unconnected. The shortage of both made ranching the only answer in many places.[30]

The absence of firewood was an insoluble domestic problem, not always revealed to the potential settler when he left the East. In some places, the only available fuel was the wind-dried droppings of animals, first those of buffalo, called 'chips'; then, of course, the wind-dried dung of cattle. But help was at hand; before 1860, the earliest timber to arrive came by steamboat, from the Mississippi–Missouri Valley. Steamers travelled along the various tributaries of the Missouri as well as the great rivers. As many settlers would not dream of paying good money for firewood, the sod came to the rescue, as did peat in woodless areas of Ireland. Most prairie sod would burn when dry enough but sod – the topsoil – was naturally the most valuable part of the virgin prairie and never as deep as peat in Ireland. Sometimes fencing, fuel and housing were all ignored and side-stepped as problems, and large spreads without livestock would be established, planted with grain one year and lying fallow for the next two or more years. These broad expanses were farmed with gangs of itinerant workers, even with hired horses or mules, which travelled round ploughing, planting and harvesting grain crops. This happened in Iowa before 1860, but by 1900 the State had the densest rural railway network in the Union, and relatively small family farms were established, with the corn-hog cycle dominant and Chicago the market-maker.

Chicago in the nineteenth century was the fastest-growing city, not just in the United States but in the world. In 1830, Chicago had a population of less than 100; in 1860, nearly 100,000; in 1890, about 1 million; in 1906, 2 million. This represented an average increase of nearly 7 per cent per annum over more than half a century, never before achieved without compulsion. All the early growth before 1870 was a function of cheap, freely available timber. At first it came from the surrounding countryside, then by lake steamer from further north and west, from Michigan and Wisconsin.

What made Chicago possible (and probably unique) was the balloon-frame construction invented by Augustine Taylor in 1833. Light but stable timber frames were bolted or nailed together on site and mounted on a prepared base; they were then clad with planking. Chimney and fireplace may have been built in place first, and the balloon-frame mounted around the core, and this later became a standard method of construction. The technique was an immediate success and was adopted worldwide, and is still in wide use today. The first Catholic church in Chicago was built like this, and the name balloon-frame was awarded by a watching priest, a cynic, who said that the church would take off in the first high wind. He was wrong, but the nearly all-wood city of more than 100,000 timber buildings was three-parts destroyed by fire in 1871, in the worst city fire since the Great Fire of London in 1666. After this, any new building in Chicago had to be clad in fireproof materials – ceramics, concrete or stone – so that sparks could not allow another fire to spread so quickly. This merely replicated what James I of England had decreed in 1612.

Timber is still the favoured material for American houses, whatever the outer sheath, and more modern buildings are still being built of timber in the United States than elsewhere. But there is another important point about timber throughout US history; cheap lumber was always the motive for transport by sea, river and canal, and then by railroad, and this brought forest products to the great cities. The real cost of housing in America was thus far lower than in Europe before 1860. Since then, the margin between the United States and the United Kingdom in particular has widened further, and today there is no place in the United States where lumber for building is not a bargain in world terms, and timber is still the preferred building material for houses, though not, obviously, for apartment blocks.

VII

The curious position of those who thought about timber and forests in the nineteenth century should be noted. American intellectuals were proud of their country's good fortune and recognized that much of the wealth of the United States could be ascribed to cheap food, cheap fuel and cheap buildings. The last two were due to huge virgin forests, more acreage per head than any nation had ever exploited before, and with a wider range of forest options than even in Canada, Russia or Scandinavia. But few pages of reflection were published before the Civil War about the rate at which the forests were disappearing, nor much about their regeneration. There was a total absence of any strategy to replace what had been consumed, often wantonly. Between 1840 and 1860, an estimated 100,000 square miles of woodlands had been destroyed by an average of 10 million people in the then West. This was in addition to any timber used after felling second or third growths elsewhere in earlier settlements. This meant that more than 2½ billion tons of wood were consumed in those twenty years, or about 12 tons per head per year.

Even more serious than the sheer amount of growing timber destroyed was its profligate use. Huge acreages were clear-felled to convert forest or savannah into agricultural land. Young trees were cut down because small-diameter trunks were easier to fell. Large trees were often ring-barked and allowed to die and were then burnt. Young trees were selected for railroad ties (sleepers in England) and this threatened future woodland, as did clear-felling, both making natural regeneration difficult to impossible in the absence of suitable mother-trees.[31]

There was no active replanting and no stimulated regeneration, as had already occurred in places in the East. There was not even an inventory made before it was too late. In the United States east of the Mississippi–Missouri and west of the Appalachians, the forested area diminished between 1840 and 1860, from 50 per cent to barely 25 per cent. The wooded area lost was equivalent in size to West Virginia, Tennessee and Indiana put together – or, in European terms, to England, Scotland, Wales and Belgium. Never before had there been such a rapid man-made change in the surface vegetation of such a large area of the Earth.

The destruction of forest between the Appalachians and the Mississippi-Missouri valley has to be regretted, and not only by romantic Greens. In all, the wooded area lost was equivalent to more than 200,000 square miles, an area nearly as big as France, or the whole land area of Illinois, Michigan, Ohio and Wisconsin, or the State of Texas.

There were all sorts of adverse side-effects: one was the destruction of wildlife habitat, which led to the consequent ruin of some indigenous Amerindians. There was land erosion, watercourse pollution and abuse, along with long-term climate and weather changes – mostly immeasurable and much of which might or might not be strictly consequential. Some small dust bowls were created, though never as seriously east of the Mississippi–Missouri valleys as further south and west, but there was a loss of the many less obvious species amounting to what might be called genocide. And so forth.

But the opposite case has never been properly made. The United States is the only First World country to have been so dependent on timber for fuel so late in life. Between 1820 and 1860, the product of about 12,000 square miles was burnt every year as fuel. This is equivalent to an area as large as the land area of Massachusetts and Connecticut combined, or the whole of Belgium. An ecologically sustainable area producing this annual amount of firewood would have been at least 1.5 million square miles, or as large as India or twice the size of Alaska. Such areas were not, of course, available, but the impression is that firewood was only abandoned because the supply of timber ran out, as in England 250 years before, and coal had to be used instead. It was not only firewood that the wealth of timber provided. Between 1820 and 1860, an annual average of 3,000 square miles of woodland was cleared for farming, to provide 12,000 quarter-sections of 160 acres each year, or nearly 80 million acres in all. This was an area four times that of agricultural land in England and Wales.

At sea, as late as 1880, nearly every one of the 30,000 ships registered in the United States was of American wood, as were houses and many other buildings at that date. In America, steel ships did not overtake wooden ships in tonnage until after 1910, and wooden houses are still preferred over brick or stone. The lateness of this near-universal use of wood compared with other advanced countries should be recognized. Its effect should be agreed to be as important as the nation's dependence on wood for fuel. Only a generation elapsed between American use of timber for nearly 90 per cent of its fuel – in 1860 – and electric light and power, coal, natural gas and fuel oil in 1890.[32]

VIII

A number of early American industries producing wealth through exports to the West Indies, Britain or Europe were timber-dependent. The more obvious were fishing, whaling and shipbuilding, with a singular commerce in ship's masts.

Fishing on the Grand Banks probably started before Columbus's first voyage. The Portuguese (and perhaps the English) fished for cod, without seeing land, as early as the 1470s. The French followed at the turn of the century, and later the Spanish and Dutch. After about 1510, many ships fished the waters between modern Newfoundland and modern Nova Scotia, having moved west from the fisheries of Iceland, Greenland and the Grand Banks; they landed their catches for smoking or salting nearby. By 1525, there were eighty wooden huts on the hills above what is now St John's harbour in Newfoundland; these were used impartially by French, Portuguese or English fishermen during the season, and they worked and lived here while preparing their catches for transport to Europe.

In 1583, Sir Humphrey Gilbert took possession of the island of Newfoundland on behalf of Queen Elizabeth. By 1599, there were 200 English ships in the Newfoundland fishery, employing a claimed total of more than 5,000 men afloat in the season; in the winter, most were ashore in Old England.[33]

In 1614, Captain John Smith came North from Virginia and spent four weeks fishing off what is now Monhegan Island near present-day Portland, Maine, south-west of the Newfoundland fishery. Smith says that he caught 47,000 fish, but who counted them? If the fish weighed 2 kg each, dried, smoked or salted, the cargo would have amounted to 94 tonnes, which sounds about right, however. Smith also named New England and Plymouth and claimed that in 1624, fifty English ships were fishing off the New England coast, further south and west than the Grand Banks.

By 1636, the date of the founding of Harvard College, at least a dozen Massachusetts settlements had been established as fishing townships, and fish was exported to most European settlements in the Atlantic and the Caribbean. There had already been a dispute with the Crown about fishing licences and about English or Colonial rights to license fishing for cod. In 1639, the *Triad*, said to be the first ocean-going ship built in Boston, took a cargo of dried fish to Bilbao and Malaga, at either end of Spain, and returned with a cargo of wine, dried fruit, olive oil and wool, exchanged for dried or smoked fish.[34]

By 1670, the New England fishing fleet probably numbered more than a hundred ocean-going vessels and three times as many undecked or half-decked sailing ships, besides sailing/rowing boats carried by the fishing ships. Cod was the chief catch, though mackerel, hake, pollock, herring and sea bass were also taken. Before refrigeration became available in the 1880s, drying, smoking and salting were the only means of preserving fish to be safe to transport over long distances. Drying and smoking were carried out ashore; fish could be packed in salt on board but the high cost of salt often favoured drying or smoking.

There were three main cod products: whole sides of merchantable quality went to Southern Europe – Spain, Portugal, Italy; smaller sizes went to Madeira, the Canaries and Jamaica; then there were what were euphemistically called 'broken fish' – small, thin whole fish and heads and tails, and these went to the West Indies to reinforce stew for slaves, and were probably the only high-quality protein the slaves ever ate.

Whaling was a very old industry in Scandinavia, the Vikings having been active in the North Sea and the Arctic as early as the ninth century, to our certain knowledge, and probably earlier. Every European fishing community regarded stranded 'drift' whales as fair game and a bonus, and settlers in New England were no exception. The next stage was to go out in small boats and drive or lure whales ashore and kill the beached animals; or they were hunted offshore and taken ashore for blubber for oil and whalebone, used for many purposes in the pre-industrial age.

Whale oil became nominally essential for lamps and candles, whilst one important use for whalebone was for strong brush-bristles – no other natural raw material could rival its strength or flexibility. The changing value of whalebone was important; Greenland whalebone fell in price from £700 per ton in 1650 to between £250 and £500 in the 1700s, and down to £150 after explosive harpoons and steel-wire bristles had both been invented following the 1860s. These prices were free of inflation and have to be increased by 70–100 to match modern purchasing power.

Although mineral alternatives were found for both whale oil and whalebone in the mid-1860s, oil from the more valuable sperm whales (first hunted from New England in 1724) was found to be of more use than that of other whales for fine candles. Ordinary whale oil ceased to be the desired oil for lighting after the development of mineral oils (notably kerosene) in the 1860s and 1870s, and solidified mineral oil for candles. But whalebone was still needed, and was priced at £250 per ton immediately prior to 1914, sometimes to produce the hour-glass figure much paraded by the Gibson Girls of that era.[35]

Whales used to be common everywhere in the world's seas, but explosive harpoon-heads and steam-whalers gave the human hunter such an advantage that the danger and skill of the manual chase gave way to predictable slaughter, with known results. Most whales are now at best endangered, at worst extinct or near-extinct, though some whales of some species can live for over a century. While it lasted, the New England whaling business was an important example of a timber-based industry, even if the profits of whaling were usually less than those of the cod fishery. New England whalers went to the Davies Strait, to Baffin Island and into Hudson's Bay. Later, they scoured the South Seas in competition with ships from Dundee, Scotland or Bergen, Norway and others. The 1860s, the decade of the American Civil War (or the War between the States) was the hinge decade for the destruction of the whale family. By 1866, explosive harpoons were mounted on steam-driven whalers, but at the same time steel wire had been developed, stiff enough to make the bristles of a brush, which before then had to be made of whalebone. The first kerosene was distilled from the first crude oil produced from wells in Pennsylvania and Germany before 1861. With mineral oil for lighting and steel wire there was no ecologically important reason to go on whaling; even whalebone for corsets could be (and was) ultimately replaced by steel.

If whaling had stopped in the 1860s, many more of these

wonderful animals would still be around. As it is, perhaps nine-tenths of all whales were destroyed after the end of the Civil War in 1865. The slaughter was by then commercially unnecessary but New England crews and (by this time) whalers from California and the rest of the Pacific coast played their part in the carnage, as did the Japanese after 1870.

In retrospect, it is more than sad that these wonderful mammals should have been nearly wiped out because the use of alternatives did not appeal quickly enough to the commercial classes of the day. Before the late 1860s, whaling was largely timber-dependent, with a significant degree of risk for the hunter, but most of the species-destroying massacres were not committed by men in small wooden boats, at risk to their own lives. After steam-engines, iron hulls, steel wire and explosives had been introduced, whales were exterminated in a form of industrialized slaughter, not by ships of wood, nor by the men or methods that figure in *Moby Dick*, written in 1851.

Ocean-going ships were built in New England as early as the 1630s, more cheaply than in Europe, but the great advantage of building ships in New England was never exploited by the English Board of Admiralty. Colonies existed to produce raw materials, not finished goods – not even capital goods like ships. So Royal Navy ships continued to be built in royal dockyards – increasingly of imported materials – to the undoubted advantage of civil servants, dockyard officials and contractors who skimmed rich, high-fat cream from the contracts.

Hulls got larger and needed the finest timber to be well-found. A big ship of the 1660s was about 100 feet long, more than 30 feet in the beam and at least 16 feet in moulded depth. A century later, when *Victory* (of Trafalgar fame) was built, large ships were nearly double in each of these dimensions, whilst the largest wooden ship ever built for the Royal Navy was 240 × 60 × 30 feet. Such wooden ships could scarcely be built today; nowhere in the world is there enough suitable timber.[36]

Masts were critical. The three sections of the main mast would together be taller than the ship was long. The base of the bottom section of each mast would be embedded in the keel. Masts suffered more stress and strain than any other part of the ship and large standing tree-trunks of the highest quality were needed – up to 24 inches finished diameter in the base. In the absence of monster trees, spliced masts could be built up of shorter, smaller logs. This technique was despised when big trees were available, but inevitably adopted when such trees became scarce. The technical virtue of spliced masts, like that of much else in history, grew out of necessity; by 1800, it was claimed that compound masts were cheaper, stronger, lighter and more elastic than those made of whole tree-trunks.[37]

Some attempt can be made, by jobbing back, to assess the numbers of masts that would have been needed. It is known that in 1800 there were more than 8,000 British merchant ships capable of sailing anywhere, and perhaps 20,000 European-type ocean going ships in the world. (Of these, only about 3,000 were warships.) In 1700 (at this date 'British' included British-American) there would probably have been 5,000 square-rigged ships, of which 2,000 were British, and in 1660 it would be fair to reduce the numbers by one-third to 3,300 square-rigged ships with many fewer – say 1,000 – British.

Each square-rigged ship would have had at least six tree-trunks in the masts, to which must be added spars and yards and the bowsprit, all of them needing great trees. If twenty trees on average were needed for masts, spars and yards for each ship, then for 1,000 British ships in 1660, 20,000 trees would have been needed and, if the life of masts, yards and spars was ten years, 2,000 huge trees every year; in 1700 it would have been 4,000 every year; in 1800, 16,000. This last demand, which would imply 2 million trees on a 125-year cycle, was almost certainly incapable of fulfilment, and therefore by 1800 compound masts had to be used. (Iron masts were only available after 1860.)

Until the Dutch Wars following the English Civil War, both Royal and merchant ships were largely dependent on Baltic masts, but when the Dutch threatened to close the Baltic and did so in 1658, New England was increasingly brought into play. Although the first Transatlantic 'mast ship' had sailed as early as 1634, masts were not shipped regularly across the Atlantic until the 1650s, and after the Dutch had captured one or two, they were sailed in escorted mast convoys. Each ship carried between forty and a hundred logs, roughly trimmed but not seasoned, seasoning being the job of the royal dockyards; Eastern White Pine was the preferred timber. By 1700, it was clear that Europe could no longer supply the number, size or quality of huge trees required for both royal and merchant ships, at an estimated annual need for 8,000 for masts, spars and yards. Timber for hulls was a more diffuse matter, and oak, which was preferred for planking, and other hardwoods came from many places besides the American Colonies.[38]

Mast trees from New England were superior to Baltic alternatives, and whilst the Danish Government (then ruling Norway, a vital source of mast timber) had banned the export of trees of more than 24 inches in diameter at base, there was no size limitation on trees from America. France succeeded Holland as England's major enemy from 1689, and French penetration south from Canada was seen to be primarily aimed at British mast trees, not only to acquire masts for French ships but also to deny them to the English. Wily French diplomats claimed part of Maine, 'to make use of the trees'. Apart from timber, Maine contained a few Amerindians, fewer than further south, where the best White Pine was to be found east of the Connecticut River, between the 42nd and 44th parallels; Europeans in what is now Canada had found few trees of comparable quality. White Pine from New England was worth four times as much per cubic foot as deciduous hardwood – hemlock, chestnut or oak – from the same forests, and six times as much as other softwoods.[39]

The mast situation led to three effects, each with unintended consequences. The area that is now the State of Maine was not a colony in its own right, but subject to the overlordship of neighbouring Massachusetts, and much of the forest escaped the direct control of the Crown. Maine was also a kind of political no-man's-land, with little European settlement and few resources to exploit except for timber and furs. Even at Independence, Maine probably had a resident white population of 2–3 per cent of its neighbouring suzerain, Massachusetts, which at 340,000 had the second-highest population of any of the thirteen original States in 1776.

Maine did not become a State until 1820, at which date the resident white population was no more than 35,000, to which must be added a few thousand Indians. But for a century, there had been a stalwart self-reliance about the British in Maine, long before the idea of political independence was articulated to plague the Royal Government. The British in Maine were as keen (or keener) on freedom as the bewigged philosophers in fine silk stockings in the drawing-rooms of Boston, and in Maine there began to be talk of autonomy as early as the 1730s.

The French started to pay Indians to harass British-American loggers, at a going rate for each Anglo-American scalp delivered. Protection was then provided by the British and armed guards were hired to defend forest workers. In time, some Maine Indians sought alliance with the French, then very few in number compared with the British-Americans. The French and the Indians succeeded in making themselves more than a nuisance and sometimes threatened to make it impossible to fill the annual mast convoy. By the late 1730s, the Royal Navy had become heavily dependent upon New England mast-timber and some in London were of the opinion that the French would have to be removed from Canada by force, in order to protect New England in general and Maine in particular.

French harassment had become intolerable, but it would not last for ever. A kind of control of the French was achieved by Wolfe's victory over Montcalm in 1759, and it could be argued that the French brought the loss of Canada upon themselves by more than eighty years of provocation about timber. This analysis suggests that French Canada became British because of the Royal Navy's need for masts, not for general Imperial reasons.

Canada was made fairly securely British, and Maine was protected from French Royalist pretension only a generation before viable compound masts did away with the need for huge trees. Would anyone have bothered if the technology had been accelerated? There is a double irony in that the removal of the French threat in the North – itself largely based on the increased demand for mast timber – materially accelerated calls for self-government in New England, and this in turn led in less than a generation to an increased call for Independence.

*

Today, we live in an age of synthetics at sea as well as on land. There are not only synthetic plastic ropes, but also synthetic canvas for sail-cloth, synthetic chemical paint and anti-fouling for ships' bottoms, and synthetic resins for many other purposes. Synthetics will not rot, but before this century wood, ropes and sails could only be preserved with natural resins.[40]

Modern tar is made in closed retorts, and most tar today is derived from coal, originally as a by-product of the coal-gas process, now often as one of the many complex compounds that can be extracted chemically from coal; these include aspirin. But wood-tar, or pitch, was distilled in Ancient Egypt and Greece. Since the reign of Queen Elizabeth, wood-tar in England had been called Stockholm Tar, which explains its Baltic origin. There is proportionately more resin in the roots of a tree than in the wood of the same tree, and small quantities of tar were distilled in Scotland from the roots of Scots Pines, the timber being used for other purposes. These supplies became wholly inadequate, and such was the demand that, before 1700, 90 per cent of all tar used in the UK was imported from the Baltic. The British required more wood-tar for their ships than any other European power, because they owned more ships than anyone else, but in the early years of the eighteenth century, Baltic tar became both difficult to secure and expensive, and they felt they were being victimized by the Swedes, who controlled the Baltic trade. So a new supply had to be found.

When demand was switched from the Baltic to America, Pitch Pine or Swamp Pine became the preferred wood, and the roots yielded very well, though in the days before sealed retorts it was difficult to separate various grades during distillation. One form of tar production used the wood, not the root, of Pitch Pine (which contains more resin than most other common species) and turned the timber into charcoal, with tar as the by-product. Charcoal was always needed, largely for iron production, and the tar was run off and preserved as part of the process, as without removing resins charcoal is not pure enough to smelt iron.

The early process was crude. Pitch Pine was packed as for charcoal burning, but drains were dug at intervals to allow the pitch to run off into a collecting basin, whence it could be loaded into barrels for sale. Since Pitch Pine tends to grow on sandy soil, there must have been impurity in the tar. The heat in such a crude process cannot be controlled, so the following modern sequence could not be achieved, unless the charcoal-burners were very lucky. At 100–120°C, water with acetic acid comes off and is now rejected. It is followed at 120–230°C by creosote; the residue above 230°C is semi-solid at ambient temperatures, and makes useful asphalt. Mixed together, the resulting heavy tar was what was needed in the eighteenth century for use in preserving wood, canvas, sails and cordage of every kind. Its use was so general that British sailors became known as Jack Tars.

So successful was the new industry in America, and so productive, that by 1720 twice as much tar was produced in the American Colonies, and exported to England, as could be used in peacetime by British ship-owners. The balance was exported, and who made most of the profits? London dealers, of course, and this became yet another source of Colonial grievance.

Iron smelted in the Colonies was also cheaper than iron smelted in Europe, and this advantage would also accrue, it was thought, to English merchants and/or investors. This happy, rosy Anglocentric condition did not last long, however. As early as the 1720s, British Americans themselves began to invest in iron furnaces in several colonies, with New Jersey first in the field. Soon there were cast-iron furnaces in Virginia, Maryland and Pennsylvania as well as New Jersey, which nevertheless remained the most important iron-smelting colony until after Independence. But by 1776 there were smelting works in every colony.

In each case there had to be the essential combination of ore bodies, labour, water and timber, always a necessary conjunction for successful ironworks. Stony New England was short of good agricultural land (and iron ore) but Yankees were eager to turn their hand to every trade, so New England entrepreneurs bought pig and bar iron from other colonies and fashioned raw iron into pots and pans and hand tools of every conceivable kind. By the 1740s, American tools for household, farm and manufacturing use were, like domestic pots and pans, not only cheaper than those imported from England but of higher quality. They were cheaper because iron was cheaper – one of the advantages of less expensive timber. The finished goods were better because Colonial iron-formers were closer to their customers.

The competitiveness of American manufactures was known before the Industrial Revolution, although the causes were never analysed; what was always a surprise to Europeans was the high quality combined with value. It still is.

There was a noisy conflict over the next stage. In view of the importance of American iron in meeting British needs, the Government determined that pig and bar iron should be allowed duty-free entry from the American Colonies; this was enshrined in a Bill presented to Parliament in 1749, '. . . to Encourage . . . Iron from America . . .'. There was opposition to this from English iron manu-facturers who fashioned objects from the raw cast-iron, and from English landowners who grew trees suitable for making charcoal, the bark of the same trees going to the tanning industry. Tanners also opposed the Bill, since they were worried that tanning bark would become less easy to find and therefore more expensive.

*

By 1749, the timber trade from the North American colonies had been essential for well over a century, especially in respect of masts but also for ship's timbers, tar and iron. These naval stores played an essential part in the naval wars during the twenty-five years before the American War of Independence. It is known to some historians that the War of Independence was started by rich men and that it was the first Rich Man's Revolution, the first successful revolution in history fought by a rich colony against a less rich Home Country. Even so, the colonists managed to wear an air of deprivation and to create that air of moral grievance so indispens-able to political revolutionaries.

Since 1776, cant about the nature of the American Revolution has become an industry. There is a degree of mythology in some histories of the United States, and for other, later revolutionaries a specious role model was established – a new republic freed from wicked royalist oppressors. The Revolution has been claimed to be largely political, rather than economic, but this too is a myth since the major grievances were nearly all about money, taxes, tariffs and the creation of credit or its denial. The Rich Man's Revolution was radicalized as early as 1772, but remained for a long time in the control of moderates who both hated and feared mob rule. There was also a large minority which (illogically?) both feared the rule of the majority yet denied the right of the Crown to rule them, claiming that London had no right to tax them. John Locke's century-old cry, 'No taxation without Representation', was used with great success by Americans against the London Government.

When shooting began, relatively few were killed, and this Revolution was also unusual for another reason. Although there was savagery on both sides, the War of Independence did not consume its own children, as did most revolutions before and after 1776. The only real sufferers were the Loyalists, who largely went home or north to Canada.

The leaders of the American Revolution would have been outstanding in any society in any century. There was more civic virtue in evidence in them than in their contemporaries in power in England. British-American leaders were also far more remarkable than people in any other colony at that date, and apparently better educated.

There might have been a different Independence movement. In no other country could Benjamin Franklin – great friend of Lord Chatham, the Elder Pitt – have suggested his alternative to revolution. In its place, he said, America should become the Seat of Empire, with the Crown lodged and the King living in an Imperial American city, and with Viceroys ruling England, Wales, Ireland and Scotland. This suggestion may have been made with tongue in an old wise Philadelphian cheek, but it was an indication of the great inherent strength of Colonial America that such a vision was plausible even if thought rather whimsical. In no other European Imperial domain could such a suggestion have been made in 1770. Before 1940, Imperial America and Colonial Britain never materialized, though it is arguable that Benjamin Franklin's vision came to pass during the Second World War.

What incontrovertibly did come about was only because of the influence of timber upon history, in this case upon the history of the Thirteen Colonies. It was timber that had made America rich.

IX

There was a unique Anglo-American means of transport in the form of the clipper sailing ships built between about 1850 and 1870 before steam became more general. These were the fastest, most profitable, elegant and hard-working wooden sailing ships ever built. Increased speed did not necessarily put men or ships in greater danger.

Clippers were usually well found, well officered and better-manned than their slower, more conventional rivals. Before the clippers, a sailing ship on a long-haul ocean voyage would average about 5–7 knots at most, and regard 150 sea miles made good in a day as an achievement. The one-way journey from London or New York to Canton – about 15,000 sea miles – would take much longer than 100 days. Allowing for 'light airs' and calm, the average was not 150 sea miles per day, but more probably 75, or just over 3 knots on average. At 75 sea miles per day, the one-way voyage would take 200 days – over half a year.

Some East Indiamen almost hove-to at night, to increase the comfort of their passengers; they were also well armed and over-manned and thus able to cope with the kind of pirates who then infested the Straits of Malacca as they do now. Moving slowly, at a few knots, Indiamen were vulnerable, built for comfort, not speed. Like portly merchants, also slow-moving, the ships' length-to-breadth ratio would be 2 or 2.5:1.

The British East India Company had a monopoly of the British Far Eastern trade, including the tea trade, before 1834, while British ships benefited greatly before 1850 from the Naviga-tion Acts, only repealed in that year. There is no need to review the effects of the tea trade upon American Revolutionary ardour before Independence, nor upon the way that tea became almost un-American for a time. But by the 1830s, there was an increased American demand for tea and increased imports, all tea at that date coming from China via Canton, the only Chinese port to which any white men had legitimate access. It was not only tea that was imported from Canton: as in Europe, there was an American demand for *chinoiserie* of all kinds – porcelain, silver, ivory, textiles, and furniture – particularly lacquered furniture.[41]

A new mythology developed about tea that pleased people of fashion and proved very profitable for shippers and owners of the new, faster ships. The public was told that 'new season's tea' was tastier, more likely to infuse better and altogether more admirable than tea rendered stale because it had taken six months to arrive from Canton. Of course, before ships had been developed to do the journey in ninety to 100 days instead of 200 days, no one had heard this sales pitch. Nor is it convincing. Properly packed, tea loses no virtue, and could and can be kept, properly sealed, for many years if need be. There was a similar 'spin' about wood ships versus iron ships, the latter being said to give tea 'a bad nose'.

Clipper ships did not spring, like some Greek goddess, fully formed from the sea. There were really four stages in their evolution. The average British or American ocean-going merchant ship in the eighteenth century would have been about 75 feet in length, with a length-to-beam ratio of 2.5 or 3:1. Next, during the 1812 war, the Chesapeake Bay schooners influenced shipbuilders. Schooners, used as privateers to operate against the British, were about 100 feet long and about 25 feet in the beam, a ratio of 4:1. The later, famous *New York*, a Collins sailing ship of the Black Ball Line, built to challenge the new Transatlantic steamships in the 1840s, was larger and longer again: 152 feet long and 35 feet in the beam, a ratio of 4.3:1. Then came the American clippers, always of softwood, usually more than 200 feet long and more than 5:1 in length-to-beam ratio. One of the largest was *Flying Cloud*, 229 feet long and 41 feet in the beam, a ratio of nearly 5.6:1. There was a 'clipper bow' to slice through the sea without impeding progress by the kind of bluffness that pushed water aside; there was a 'champagne glass' stern, later known as a 'clipper stern', that was used in both sailing ships and steamers for many years.

As the Clipper Age evolved, American clippers tended to grow longer and carry more, up to 2,000 tons of cargo. Because of the shortage and high cost of wood in Britain, British clippers inclined, as time went by, to just over 200 feet in length, with a beam of about 40 feet, a length–beam ratio of 5:1. Later British clippers were usually of composite or mixed construction with wooden planking on an iron frame.[42]

More length in the waterline length of a sailing ship's hull produces greater speed, other factors being equal, but neither a V-bottom nor a flat bottom has ever been proved to be significantly faster. Obviously a slab-sided hull with a flat bottom is much easier to load and unload, and the loading of a high-value, relatively light, very vulnerable cargo like tea was a work of art, made easier in a flat-bottomed ship. So the flat-bottomed clipper won the day, nor could anyone prove then or now that any other sort of ship's section was faster than a simple, square U – safest, too, when a ship goes aground.

The key to the clipper's success was that no early steamship, sailing round either the Cape of Good Hope or Cape Horn to China, from London or New York respectively, could carry enough coal, even without much of a cargo, to cover more than 3,000 miles without refuelling. With 1,000 tons of cargo, the limit would have been only 1,000 miles. The 15,000 miles out and back would have required many stops at coaling stations, which did not then exist. There were also the unfavourable economics of the early coal/speed ratio.

It was in the China trade, before the Suez Canal was opened, or in trips to Australasia or voyages across the Pacific, that sail survived longest. In the end, it was the Suez (1869) and the Panama (1914) Canals, along with the provision of dozens of coaling stations worldwide, that killed the sailing ship; but the balance was long in doubt before triple-expansion engines, steam turbines and more efficient boilers were developed.

In fact, though clippers no longer hauled tea from the mid-1870s, many sailing ships survived until 1914. One reason was that speed in steamships was expensive. In 1900, a 10,000-ton Transatlantic liner with triple-expansion engines burned 1,050 tons more coal across the Atlantic if it travelled 3 knots faster – 17 knots instead of 14. Put another way, to travel 3 knots more slowly was to earn $25,000 ($2 million today) more per crossing from extra freight or passengers carried and less coal burnt. This had to be balanced by passengers' demand for speed in service, the 3 knots' extra speed saving two days between Liverpool and New York.

After 1900, the steam turbine changed the economics to a certain extent, but speed always cost money, as did, the argument ran, the time speed saved.

*

Right at the end of the clipper's short period of glory, in the very year in which the doomed Empress Eugenie staged the opening of the Suez Canal, the British built a famous clipper ship, and one that still exists – *Cutty Sark*.

She was not the first composite-built ship, just the most famous. She had an iron skeleton-frame, fixed into a wooden keel of the best elm wood. The iron frame was clad in teak planks 6 inches thick, and covered below the waterline by copper sheathing to prevent marine growth; this copper was later to be replaced by anti-fouling paint. In *Cutty Sark*'s day copper anti-fouling sheet made wooden planking essential, even if laid on top of iron plates, because of the electrolytic action between copper and iron.

The iron skeleton and the teak planking imparted a stiffness to the structure of the hull that was apparently absent in American softwood clippers, and was almost certainly responsible for the long, successful career of *Cutty Sark*. All-softwood clippers rarely sailed as well after their first two or three trips, probably because their wooden frames and pine planking moved fractionally and adversely changed the shape of the hull. Softwood hulls were too flexible in the strong winds necessary for high speeds, and of course the same winds also generated rough seas.

Cutty Sark was of 973 registered tons, built in Aberdeen, Scotland, for £17 ($85) a ton, or £16,541 ($82,705) for the whole ship. This was a bargain, £2 ($10) a ton cheaper than a British all-wood ship. She had 10 miles of standing and running rigging, and 32,000 square feet of sails. In Force 7 winds these developed the equivalent of 3,000 hp, driving the ship at 17½ knots, more than most steam-ships of the day could achieve.

Cutty Sark made good, many times, more than 360 sea miles in twenty-four hours, some days an average of 15 knots, and occasionally nearly 400 sea miles (more than 16½ knots). But her worst days were far inferior to any steamship's – no sea miles, 20 sea miles, 40 sea miles, 60 sea miles – and it was the unpredictability and sheer unreliability of wind (and therefore of sailing-ship progress) that made for the gradual triumph of steam. After the Suez Canal was opened, steamships captured the tea trade and reduced freight rates from £6–7 per ton of tea to £3.50, and then £3 per ton, and passage times to a regular, non-racing, non-sporting ninety days. No more was heard of the need for the tea clippers' extra speed or the benefits of a wooden hull to keep tea sweet, and these were shown to be mere marketing ploys.

*

After 1850, American clippers plied another trade besides tea, which also required high speed but was rated as an internal US 'coasting' voyage as far as the law was concerned, so was legal for American ships only. This was after gold was discovered in California.

There were three ways to reach California between 1849, when gold was discovered, and 1869, when the Transcontinental railroad was completed. One was by rail and then covered wagon across the Continent, which was infested west of the Mississippi by Indians who were often, and reasonably, unfriendly. Wagon routes were not always close to the essential forage, and sometimes not near enough to the even more essential water; wagon travel was far too expensive for freight of any kind. But many would-be settlers preferred the land routes.[43]

Others, more adventurous, might take a steamer to Panama, cross the Isthmus, and board another ship to San Francisco. By 1855 the Panama railroad had reduced the time on the crossing of the Isthmus to less than a day. But before 1905, the route had the serious handicap of three killer diseases: cholera, yellow fever and *Falciparium* malaria. Only courageous, wealthy people in a hurry went by Panama; the double trans-shipment was too expensive for goods, as labourers did not survive long enough or were not fit enough after a while to work well in the disease-ridden climate.

So the voyage round Cape Horn became the order of the day for most goods and for those people without the requisite time and/or high spirits to go by wagon, or the money and/or courage to risk the diseases of Panama. Many passengers travelled to San Francisco via Cape Horn, even if the trip cost only $300 steerage by Panama, and most freight, and all valuable freight before 1869, went to California by clipper, round the Horn.

With almost every vigorous worker in California in the gold fields (when not in the saloons and brothels), the humblest, most pedestrian everyday goods made three, four, five times their New York prices when sold in San Francisco, and shipowners and shippers made fortunes.

Wheat, flour, bacon, ham, beans, whisky, textiles of all kinds, boots and shoes, and especially picks, spades and shovels for the digging in the gold fields – all these were in great demand. A cargo of the most basic type of household goods that people in the East bought every day of the week would be worth in California more than the value of the clipper ship that had brought the cargo round the Horn.

This was just as well for the owners of ships, because at one time there were nearly 200 sailing ships abandoned by their crews in San Francisco harbour. Crews would start jumping over the side as soon as they entered San Francisco Bay, and at least one ship's master was left with only the harbour pilot to help him drop anchor. Ships' captains had to go back East by land, unable to find crews or another ship to return by sea. Abandoned ships, even fine clippers, were often sold to be broken up for wood for housing and other buildings, most of which burned down in the earthquake and fire of 1906.

Ships' officers were also known to desert their captains, and the only consolation and compensation for their owners was that, when they jumped ship, such crews were never paid. In this random way California gained the most extraordinary collection of male immigrants drawn from every country on earth. Nor were the numbers negligible; men who jumped ship over the three years 1848, 1849 and 1850 are said to have numbered 10,000. The total population in California at the 1850 census was only 90,000, so sailors who had deserted might have amounted to more than 10 per cent of the new State's total population.

Both the Suez Canal and the Union Pacific were opened in 1869; the clippers lost both the tea traffic and the Californian business, so they engaged in other long-distance trades, one of which was carrying wool from Australia to Europe. More dangerous cargoes, like coal or guano, were much more likely to be carried by windjammers, which, unlike clippers, were of steel. Because their hulls would not burn, they were much safer for cargoes that ignited spontaneously. An iron or steel ship had a chance, given a brave crew and sufficient pump capacity, even if holds ignited either spontaneously or otherwise. Fire in a wooden ship meant that vessel, crew and cargo were almost certainly doomed, and fire was a common cause of loss.

After 1880, insurers increased premiums so much that cargoes prone to spontaneous combustion were virtually uninsurable in a wooden-hulled ship. The option open to owners was to take the risk or refuse the cargo. Windjammers therefore became the ships of choice over long distances.

In 1830, an old-fashioned bluff wooden East Indiaman had a crew of sixty, a cargo capacity of about 1,000 tons, and an average day's gain of less than 100 sea miles. Fifty years later, a full-rigged ship with a cargo capacity of 5,000 tons was designed to be worked by a crew of only thirty and would march across the oceans at twice the effective speed of the Indiaman. Productivity per man had therefore increased by nearly ten times in fifty years, without the cost of steam; windjammers were consequently very effective in the market-place. One of the best recorded voyages was completely round the world in 153 sailing-days, from London via Algoa Bay, Lyttleton in New Zealand, and back to London, carrying a different cargo at each stage. No steamer at that date could have competed with such a performance. If the time had been equalled, the cost would have been far greater; if the cost had been matched, the voyage would have taken much longer.

But it was steel, not iron, that defeated wood as a material. In 1860, more than 60 per cent of the tonnage launched in the United Kingdom was still built of wood or composite material. In 1890, after economically produced steel had become freely available, less than 1 per cent of the UK tonnage launched was of wood or composite construction, less than 4 per cent of iron, and more than 95 per cent of steel construction.

In 1860, nearly 90 per cent of the world's tonnage at sea was still of wood or composite construction; under the US flag, more than 97 per cent; under the Union Jack, rather less than 80 per cent. Iron ships represented the balance as steel was not then in use. In 1890, though wood and composite construction were still responsible for nearly half the tonnage owned worldwide, the proportion in the United Kingdom was very small – less than 5 per cent – while in the United States wood was still the material in the case of over 90 per cent of American-owned ships. So wood continued to be an essential material for ships in the United States long after the Civil War.[44]

X

Thus steam-power in Britain, and Britain's original lead over every commercial rival, owes its origin to the timber shortage that was obvious by about 1600, grave by 1700 and structural by 1776.

It follows that the Industrial Revolution occurred first in Britain for one overriding reason. The chronic shortage of timber led to the early use of coal as an alternative fuel. In turn this led to the need for steam-pumps to deal with water before deep coal could be mined. Consequently this meant a head-start for coal, iron and steam of at least fifty years compared with any other country in the world.

Equally, England's naval supremacy at the time of Nelson can be claimed to have stemmed in part from the unsuspected chemical accident involving wood 250 years before, in the time of Henry VIII. And it is arguable that the British Industrial Revolution, dependent as it was on international trade, would never have occurred as it did if the Royal Navy had ever been seriously defeated by the Spanish, Dutch or French. One factor which prevented such a defeat was the great virtue of English ordnance. The prime cause was that high-phosphorous Sussex ore contained trace elements which prevented crystalline faults developing during the cooling process after casting. This cooling process was, of course, uncontrolled and unappreciated at the time, and the process of casting, cooling and crystallization entirely arcane. This chemical advantage continued until iron gave way to steel for ordnance in the 1870s.

As everyone knows, the Industrial Revolution began in England before anywhere else, and before any considerable quantity of iron was smelted with coal. Yet England raised and burnt more coal in 1500, 1600, 1700, 1800 and even in 1850 than all the rest of the world put together. And the Industrial Revolution could only have begun as it did in England because the other uses of coal, unconnected with iron or steel, released wood for what were then the two essentials – the making of charcoal for iron, and the building of ships. Until the late 1700s, there was no substitute for wood for smelting iron, and when coking coal was introduced especially for ships' cannon, England's lead continued to be important for more than fifty years.

The English timber shortage, like the shortage which earlier afflicted Venice, for different reasons, could have altered national policy. By 1776 the English were fortunate enough to have coal as a substitute for wood for all purposes other than building ships. But they had also found the wooded colonies in North America essential to national survival, and refused to be defeated, either by circumstance or by their Continental enemies. When they *were* defeated, it would only be by their own kith and kin – the American War of Independence was the only conflict that the British lost in the eighteenth century. But the defeat mattered little as far as trade was concerned – the new country and the old remaining in close concord as far as business was concerned.

Economically, Independence made little difference. The new United States and Britain remained each other's best trading partner, and the British investment in the United States, and vice versa, continued to be most important and remains so to the present day. The cultural exchange is equally important in both countries, something that many Continental Europeans find hard to understand, especially the French, who bizarrely believe the Channel to be narrower than the Atlantic.

XI

By 1850, 65 per cent of US workers were still employed in rural areas, 35 per cent in urban centres. In the United States, it would not be until after 1920 that the rural/urban balance matched the same 50:50 British ratio reached in 1851. In other words, the British, the first industrial nation, were seventy years ahead of the Americans in their urbanization – the sometimes almost compulsory movement of people into urban areas.

The most remarkable effect of the timber bounty on the United States must be the belief in personal opportunity that most European-Americans have felt throughout their history. America has been a land of opportunity not only for the oppressed and the underprivileged, for the deprived and impoverished of Europe, but also for every nineteenth-century European-American. The single most obvious contribution to this feeling of opportunity-for-all was the Frontier.

The adventurous, the mavericks, the misfits, the immature, the bold and the daring could all take advantage of the Frontier. Even those who never left the East benefited from it. In times of hardship in the East, the courageous and the enterprising might move West to improve their fortune, thereby leaving more chances for those who stayed behind, however unintentional such relief might be. It was notable in New England that as many of the original, more adventurous, Yankees left for the West, the remaining stay-at-homes developed more staid characters. Notable, too, was the westward movement of New England forest-workers into areas like Michigan or Minnesota.

The American Frontier's importance in history is in turn impossible to understand without timber. Imagine a geographical hypothesis – a great treeless plain that began in Western Virginia and Pennsylvania, on the western slopes of the Appalachians. Without steamships or railroads, themselves impossible to fuel without wood, and with little indigenous native edible life and few furs to catch and sell, no timber to build houses, fence farms or make fires to keep warm or to cook food, settlers would have had to rely entirely on what was imported in unreliable and expensive wagons.

There is a more general social and human point about wood as fuel. As early as 1860, in England and Wales, 60 million tons of coal were mined annually and more than 250,000 men employed below ground in coal production and as many again in distribution. The same thermal value in the wood-fuelled United States in 1860 required about 150 million cords of wood, the product of a little more than 2 million acres of virgin forest. At a modest estimate of 1,000 cords per man per year, this would have required less than 150,000 man-years. In England, fuel required men to work deep in dark coal mines, usually with long, narrow seams of coal. In the United States they worked in the open – in forest or savannah.

What would a manual worker rather do in 1860: dig coal or cut and cord wood? Which manual occupation, without any mechanical help, would be more likely to radicalize workers? Which employment would be more likely to result in high accident-rates? In which job was a man more his own master? The obvious answers to these questions illustrate the great social benefit to European-Americans of the American forests.

Both Russia and America enjoyed huge areas of usually unin-habited, forested land in the nineteenth century. In the Russian East, most of the settlers were under compulsion of one sort or another; in the United States, the West was for free men. The Frontier was not only a place but also an idea, and an idea that altered its shape as did the geography over time. It was an idea made palpable by the existence of the wealth of timber with which the early antebellum Frontier was covered. Timber, apart from all the material advantages it brought settlers from as early on as the *Mayflower*, also made human aspirations possible.

The unique American idea of the Frontier was therefore made tangible only because of the wealth of timber. And without the wooded Frontier and its settlement, the United States would have been a very different country, probably never the world's leading economy, and almost certainly unable to win at least one World War as well as a Cold War, with all that has meant for the world in general and for Europe in particular.

EPILOGUE

Some sceptics have rubbished the claim that global warming is caused by burning fossil fuels to excess. Others have blamed global warming for the decline in the area of the world's forests. Yet others have put the claim about the excess burning of fossil fuels firmly in the politically correct category, where it joins the irrational fear of genetic modification and an equally unreasoned adulation of organic food. This sort of sceptic wants to know why the grafting of fruit trees is not genetic modification and what, in all that is sacred about the use of language, is food that is inorganic.

Other, more scientific doubters follow the rational process of sceptically reviewing the evidence. The rational support for scepticism about global warming is fourfold.

First, combustion efficiency is an important factor. There are more than ten times as many combustion devices (internal and external) at work in Europe as there were fifty years ago, yet they each emit much less pollution because combustion is now so much more efficient. In comparison, and this is very relevant to the Third World, burning wood on an open fire produces more than twenty times as much greehouse gas as does the production of the same net energy when generated by a modern gas appliance fed by natural gas.

Second, there are huge changes in world temperature in history. In 1200, Europe and North America were at least 2–3°C warmer than they are today; grain was grown in Greenland, good wine made in Northern Europe, including England. More significantly, Rome was, at the time of the millenium of 1000, notably cooler than in 1200. Since 1000 there have been two major periods of warming and cooling, as well as climate change over shorter periods of 30–50 years, and some even more frequent cycles. When peaks or troughs coincide, even greater changes take place, but none of the changes was connected, before the 1960s, with carbon burn, however much the scope of the variation.

Third, it is arguable that US forest depletion since 1800 has been a more substantial cause of global warming than the carbon burn that has multiplied by 1,500 times in the US since 1800. (The population is sixty times what it was and the individual standard of fuel usage is 25 times what it was in 1800. Trees cover only about one quarter of the area they clothed in 1800.) Doubt has been cast upon the efficacy of growing woodland as 'carbon sinks'. But forests, old and new, may offer an alternative to limiting carbon burn, which would, of course, restrict economic growth except where there can be great improvements in combustion efficiency. This factor is especially important for the 40 per cent plus of the world's people who in the year 2000 had no fuel apart from wood, nor clean water, nor effective sewage systems, nor much hope of material progress. To electrify this huge population – about 2,500 million people – preferably using renewable sources of energy, in exchange for good land management, would not be an impossible expense for the rich nations. This would meet much of the theoretical need to reduce the excess carbon dioxide created by increased fuel burn. But it would also be frankly and (perhaps unacceptably) 'neo-colonialist'.

Finally, there is a neglected factor in the arithmetic of carbon and carbon dioxide. This is the effect of rainfall, which is beneficial, and not only for the obvious reason that without rainfall there is very little plant growth. It has hitherto been believed that carbon dioxide can only be 'sunk' by plant growth. But it is now held by many authorities that rainfall on the western side of the Rockies, for example, may sink, by direct action, as much carbon dioxide as do the trees growing in the same place and nurtured by the same rainfall.

What the direct effect of rainfall is worth in terms of carbon-sink, how much should be attributed to each tonne of rain falling and what each tonne does to the carbon dioxide it meets is not, at the moment of writing, quantified. But the rainfall factor can be held to be directly significant and it has been, to date, virtually ignored in almost every calculation about these matters.

Notes

1. See the effect on Virginia of the Stuart ban on the growing of tobacco in the United Kingdom (page xii).
2. John Evelyn (1620–1706), a Renaissance man born in early modern times, helped found the Royal Society and was an authority on architecture, gardening and trees, and a great diarist. His *Sylva* was published in 1664.
3. The cost saving of iron compared with bronze was 10–1 in the case of cannon. Sussex cannon in 1550 cost £10 per ton (equal to only £1,000 today) and shot was less than half the cost by weight, and reusable, while stone balls were not.
4. Iron was first smelted with coke by Abraham Darby in 1709, at Coalbrook-

dale in Shropshire, England, and in France in 1781 and in Prussia by 1789. The problem with coke-derived cast iron was the level of impurity, which made the production of wrought iron impossible until Henry Cort (1740–1800) invented 'puddling' to remove the contaminants. (The demand for wrought iron in 1770 was five times that for cast iron.) In the United States, the use of anthracite to make cast iron became important after 1850, after the development of the hot-air blast in place of the ambient-temperature air previously used. But wrought iron was still more cheaply produced by smelting with wood, except in England.

5. The word 'smog' (smoke + fog) was only coined in 1905 in London, but the condition was certainly known in that city more than 500 years previously. However, it must have been smoggy on more days in 1600, in Shakespeare's time, than in 1400, the year that Geoffrey Chaucer died.

6. Smoked as opposed to salted fish need the right sort of wood and to be competitive, the wood had to be found at the right price, locally. In this trade, the Americans had the advantage.

7. See the works of Samuel Buck (1696–1779) whose views of English towns, sometimes collaborating with his brother Nathaniel, nearly always include kilns for nearby building developments.

8. True porcelain, produced in China before AD 400, was not successfully made in Europe until the first half of the eighteenth century.

9. See Sir John Clapham, An Economic History of Britain, Cambridge University Press.

10. Poor Laws, intended to relieve the misery of poverty and prevent vagabondage, often tended to exacerbate problems instead of solving them. The Poor Laws existed in England from 1552 until 1834, providing varying forms of relief, indoor and outdoor, intended to mitigate poverty without reducing the incentive to work. This task has proved very challenging for rulers throughout the centuries, and solutions are still not easily discovered.

11. Though later important in coal measures, these early pumps were first used in Devon and Cornwall in tin and copper mines.

12. This connection between steam-pumps and coal – a new, prime source of energy – is historically almost wholly neglected.

13. James Watt (1736–1819) used an external condenser, which meant that the cylinder did not lose its heat between strokes; this conservation of energy made Watt's engines much more efficient. Horse-power also rose in seventy-five years from 4.3 to 40, or about nine times.

14. UK use of non-human power in 1800 was at least 40 per cent of world use, and at least five times that of France, which had a population three times as great. So each Briton was assisted by fifteen times as much power as each Frenchman. This is one reason why Napoleon lost.

15. See Asa Briggs, The Power of Steam, London, 1982.

16. See L. C. Gray, The History of Agriculture in the Southern States to 1860, Washington, DC, 1933.

17. Barbed wire was developed in the United States in the 1870s to replace ineffective smooth iron wire which cattle broke by rubbing. Steel barbed wire,

galvanized within a decade, became cheaper than plain iron wire by 1890. Five tons were made in the United States in 1874, 200,000 tons in 1900. (It later became essential for the protection of trenches in the First World War but not in the Boer War.) Wire-netting for the restraint of smaller animals such as sheep, pigs, even chickens (and tennis balls) came later, with only 50,000 tons made in the United States in 1900. In Britain, the Barbed Wire Act of 1893 nominally prohibited its potentially dangerous use alongside a road. This has been almost universally ignored, as any traveller on British roads can testify.

18. The term 'pork barrel' has a special connotation in the United States. In the South, before 1860, barrels of salt pork were breached and their contents freely offered to slaves on Feast Days. Later, in the 1870s and 1880s, the expression was used by disgusted commentators to denounce the practice whereby elected officials used public money to reward their constituents or, worse, those who had subscribed to their campaign expenses. The practice is widespread in every 'democracy' and as prevalent in Europe as in the United States. (In other parts of the world, the rewards usually go directly to members of the ruling party.) There is no cure for this dishonesty as long as those who make the laws benefit from this form of corruption. In Britain, every effort to prevent 'pork barrelling' has been defeated by the self-interest of politicians.

19. Thanksgiving Day, now the fourth Thursday in November, derives from the Day of Prayer set aside by the Pilgrims in Massachusetts after their first harvest in 1621. It is believed that the Pilgrims were enabled to survive until that date by the succour of friendly Amerindians.

20. Maize (after the Indian *mais*) is called 'corn' in the United States because all native staple cereals were also called 'corn' in Britain: wheat in favourable places, oats in the wet West and in Wales, rye on poor thin land like the Brecklands of East Anglia. When the Pilgrims found maize in use as a native staple, they called it 'Indian corn', then 'corn'. It is not, by the way, a cereal.

21. There was not much wild maize where buffalo or elk were thick on the ground.

22. Some Pilgrims gave thanks that 'The Good Lord sent the Plague to save us from our enemies'. There is new research suggesting that Amerindian populations were much reduced by disease *before* the white man arrived, certainly in New England and even in the Amazon region.

23. Chickens and turkeys share Fowl Pest, a lethal disease not recognized before the nineteenth century. Were there other, earlier afflictions?

24. The United States was singularly free from Luddites, said to be named after Ned Ludd, a weak-minded Englishman who smashed two stocking-frames in 1779. Later, self-styled 'Luddites' destroyed other textile machinery in the harsh times after 1811. They were protecting their livelihoods, they thought, from new technology. Modern Luddites exist in every human activity, and are as common in the professions as among blue-collar workers.

25. Robert Fulton (1765–1815), an Irish-American born in Pennsylvania, was a self-taught engineer.

26. On the Erie Canal, the barges were huge by British narrow-boat standards: the beam was nearly 18 feet and the length nearly 100 feet.

27. 'Watering' stock meant issuing far more shares than were represented by physical assets. Thus a negotiator might receive an introduction fee in the form of shares in a new railroad, basically for being in the right place at the right time.

28. Which came first? The British desire to be stand-offish, or the small compartment in which privacy was more easily maintained?

29. Some modern research suggests that inland Native American populations were drastically reduced by European diseases before Amerindians actually met large numbers of Europeans. This is generally known about some parts of New England, but the novelty is to ascribe the same syndrome to inland North America. It is said, for example, that denser populations of Amerindians cleared large areas of the previously wooded Great Plains by fire to make 'buffalo parks' to provide easily hunted meat supplies. Other areas were without forests because they had been cleared by fire to allow for primitive agriculture.

If these theories are correct, then many forests in the British-American Colonies settled later than the 1600s should be dated to a point, post-*Mayflower*, when enough Native Americans had been killed off by disease to permit forests to regenerate. In the case of buffalo lands, the beasts themselves, if numerous enough, would prevent regeneration. In the absence of great numbers of elk or buffalo, new woodland would be the pattern.

These theories are not generally proven, but they fit some of the topographical facts well enough to make them an attractive possibility for those inclined to believe that North America was not an 'empty country' and that white invaders were unwittingly guilty of biological imperialism, not only in New England.

30. 'Heft' has several meanings in the dictionaries. One, not present, is the local word in use in the North of England and in Scotland, applied to animals. 'Hefted' means that sheep and cattle have recognized their home as a particular unfenced area and normally stay within that area. The same valuable characteristic was treasured in animals in the American West.

31. If natural regeneration is desired – and the technique is now widely used by good foresters – a fine 'mother-tree' must be left to re-establish the forest. Depending on how seeds are dispersed, one mother-tree must be left in any given area. Clear-felling obviously delays regeneration for many years.

32. In 1860, coal supplied only 5 per cent of US non-animal fuel needs and oil a negligible proportion. In 1890, there were 600 oil or natural gas companies incorporated – not, of course, all in production, but the number is indicative.

33. Some writers said that cod could be caught in a bucket in many areas from the Grand Banks westward, before 1650.

34. Olive oil? The reader may well blink, but the *Triad* returned with this Mediterranean luxury to New England on her return voyage in 1639–40.

35. Gibson Girls had fabulous figures: 34–19–32 was typical, a 16/17-inch waist possible. In real life, the corset often provided the pinched-in waist and the pushed-up bosom. After a substantial meal, or at bedtime, the corset was removed with a sigh of relief. For a long time whalebone was kinder than thin steel strips, which are generally used today. It is probable that metallurgy is much improved since Edwardian times.

36. HMS *Victory* was most recently (1998) repaired with tropical hardwoods,

because no European or New World temperate woods could today provide trees large enough for the 'knees' on the turn of the bilges. But HMS *Victory*, launched in 1769, was certainly at least half – perhaps three-quarters – built of American timber, since there was a grave shortage of English timber for the job. Hearts of Oak, indeed.

37. The French Navy, unable to find masts as good as those in Maine, used compound masts as early as the 1750s, but for English views see the difference between the entry in the *Encyclopaedia Britannica* of 1740 and the edition of 1789. The latter tended to approve of compound masts, the former did not.

38. A warship's planks had to absorb the shock of an enemy's broadside, but also without much splintering. Splinters, which could move very fast, caused terrible casualties. Damp or wet oak splinters less than other hardwoods.

39. After the US Declaration of Independence in 1776, the Royal Navy tried to source their timber needs in (British) Nova Scotia, but the wood was not of the same quality as that from New England. Imports from Maine were resumed after peace in 1783, and the ships that defeated Napoleon were largely built of American wood.

40. Greens should note that their preferred activity – recycling – used to be called 'rotting'.

41. The effects of the tea trade upon European taste are described in *Seeds of Change*, London, 2002.

42. The Royal Navy in the war of 1812 discredited US naval ships as 'Fir Frigates' whose lines altered adversely after a few months' hard sailing. Despite this disparagement, US frigates had the best of the fighting in that war.

43. In the same twenty years before the railroad was completed in 1869, the Pony Express played a legendary role in linking the railheads at either end. Light, valuable goods and important mail were carried by a succession of riders and horses fast enough, in theory, to evade hostile Indians.

44. The upper limit on the length of a wooden ship was 280–300 feet, above which it was not possible to achieve linear strength, since even the massive keel had to be in short sections. A long wooden ship was also less watertight than a short one. Iron ships, built from the 1840s in Britain, had a temporary triumph, giving way to steel vessels by the 1880s.

WINE

The Grape's
Bid for Immortality

I

Once upon a time, probably between 10,000 and 12,000 years ago, a hunter-gatherer found some grapes that were sweet to eat immediately, when picked ripe from a particular vine. Other grapes, on another vine, were improved by being allowed to dry on the vine before they were gathered and eaten (or stored) and were valued for their sweetness. There was a third vine, with different grapes, which, regardless of how long they were left to ripen, were never as good to eat as the first two varieties; they tasted sour and were probably neglected on that account, except by small boys of the kind who nowadays pilfer and eat sour cider apples or other unripe fruit.

The first vine was producing what we would now call *table grapes*; the second generated grapes suitable for *currants* or *raisins*, and the third produced grapes best for making wine.

Several millennia later, after mankind had produced containers of wood or pottery that held liquids without taint or leakage, successful wine-making began. Although the poet John Milton wrote that there were wine grapes growing in the Garden of Eden, it is difficult to believe that wine-making could have started before pottery containers or tools good enough to make wooden containers were invented. It is known that goatskins were used very early on as vessels to transport liquid, but it is perverse to believe that grape-juice stored in a goatskin would ferment properly and become a sound wine. The archaeological sites on which grape-pips and other evidence of ancient wine-making have been discovered also reveal the essential pottery shards, always an early indicator that men and women had ceased to be hunter-gatherers and had settled down at the beginning of the long road to the present.

Early wine-making sites tend to be close to the Caucasus Mountains, a range that most authorities believe to be the home of the wine-grape. From Armenia, the art of making wine (and growing the vines themselves) passed to the Fertile Crescent and to Egypt. From the coast of the eastern Mediterranean, wine-making spread to the islands, notably Cyprus and Crete, which were early homes of other forms of higher civilization. This pattern of dispersal is more than speculative, but there are positive indications of viniculture in both Ancient Iraq and Egypt before 2000 BC, even though neither country is considered to produce great wine today. People were probably less fussy 4,000 years ago, but it is notable that in Egypt growing wine-grapes was profitable and was treated as a State monopoly. The Pharaonic State was dominated by priests, bureaucrats and soothsayers, and there were apparently good reasons why the wine industry, which is today so obviously favoured by individuals, should in those days have been controlled by the State.

Several hundred years later, about 1500 BC, after Egypt conquered what is now Syria and Lebanon, amongst the spoils of victory were quantities of wine-vines, plus a great deal of the famous local cedarwood in which Egypt was deficient, along with most other good timber. There are, from this date and from earlier times, complete drawings, almost in strip-form, in Egyptian tombs, illustrating the whole process of wine-making. These pictures portray a scene that did not much alter in any essential in France or Italy until the 1950s, though vats for fermenting grapes were smaller in ancient times. According to Homer, the Mycenean Greeks originally used wine to help both victims and priests in the ceremony of sacrifice to the Gods, but later, as wine became more widely available, it was drunk, as it is today, in the home.[1]

Later, in pre-Attic Greece, wine was imported from Lebanon, landed at Athens, then the vines themselves were imported and by 500 BC Greece was producing wine in quantity and exporting it to

Egypt and other places less suitable for viniculture. In Egypt, prior to the arrival of the Hellenic Greeks – just before 300 BC – beer made from wheat, or 'wine' made from palms or dates or honey were the popular native drinks. These were presumably for the less well-off, since wine from grapes has always been more difficult to make and to keep and therefore more expensive and more sought-after than more easily-made alcoholic drinks. Nor was wine usually drunk only as a source of alcohol, pure and simple, since alcoholic drink made from cereals (or even fruit) is more easily made and therefore cheaper.

The Ptolemys, the Greeks who ruled Egypt from about 300 BC, demanded better wine, so vines were imported, as cuttings, from all over the eastern Mediterranean and planted in the Nile Valley. The Ptolemys also taxed wine, claiming as much as half its retail value; this was a wise move because it is always effective to tax either an essential (or a non-essential that becomes an addiction) – witness the modern revenue-raising potential of tobacco, alcohol or oil products.

A basic point about wine that few can appreciate today is that much of its ancient virtue was due to the unreliability of water as a thirst-quencher. Although the causal connection between enteric disease, the alimentary canal and the seepage of human sewage into drinking water was unknown until just over a century ago, it had long been recognized that unfamiliar water was unsafe. In a village, the inhabitants might share their microbes and have a common resistance to local germs, but village micro-organisms would create havoc in the alimentary system of an innocent stranger. The reverse would be true if the villagers travelled.

So a stranger arriving in a village would be unwise to drink water unless it was drawn from a clear running stream remote from any dwelling. The rule was to drink wine or beer since the actual process of making either kills most germs. Ancient populations did not have a great deal of scientific imagination, however. Because wine was safe, though no one knew how or why, it was also thought that adding wine to water made water safe. This is not true, since no wine – however alcoholic – has as much disinfecting power as the process that made it.[2]

We are traditionally taught that Greco-Roman civilization was based on the great Mediterranean triad – wheat, the olive and the grape. This is only partially correct, and the triad was subject to various technical stages. First, iron needed to replace bronze as an everyday metal, iron being better (as well as cheaper) for implements for ploughing and other cultivation. But iron for farm implements also implied iron for weapons of war – if you had an iron ploughshare, your neighbour might have an iron sword. So the new Iron Age was notable in places for a disturbing ebb and flow of migrant peoples armed with iron weapons of war, conquering lands inhabited by peaceful peasants who used iron to make implements to cultivate the land for growing cereals.[3]

Second, the increase in populations that followed the early use of iron in many places in the Mediterranean led to excessive depletion of forests, and this led in turn to the erosion of hillsides.

A point was reached in many places where there might be little depth of soil and no hope of growing cereals on slopes that had once been used for grain. On hills around Athens, many former grain fields were now planted with olives or vines, which were much better at preventing erosion, because their roots went far deeper and were permanent, unlike the roots of cereals that die every year. But few Ancients considered the problem of erosion as much as the virtue of the wine-vine, which in poor surface soil conditions drove its roots down deep in search of food and water, picking up those micro-nutrients that give good wine its character in the process.

It is possible that the Greeks of Periclean Athens (c. 500 BC) learned that the best wine comes from the poorest soil. On the other hand, the use of resin to preserve wine, as in retsina, a practice we know was in place by 400 BC, suggests that the idea of fine wine was not really part of Athenian culture.

Nor was bread-wheat at first. There is ample evidence that the great land victory of Marathon (490 BC) and the sea victory of Salamis (480 BC) were fought by Greeks who lived on porridge. This was probably made from barley, to which sometimes – but not always – a fermented (sweet) mixture of fish was added. There was no sugar grown in the Mediterranean at that date, so honey would have to serve as a sweetener. Bread-wheat only became a staple for the Greeks after they had colonized the land of the Black

Earth and they probably first used bruised wheat for porridge and then discovered the virtues of Black Earth wheat for risen bread.

The porridge diet obviously did no harm to Greek courage and physical prowess or to the high intelligence always associated with Athens. Athenians were keen traders, and discovered soon after they had become Empire-builders that their own poor land was far more effective and valuable for producing wine for export than for growing cereals for porridge or bread. So wine became an export commodity, and the ratio of the gross product of a given area in vines to the same in cereals could have been as high as 20:1. In the alluvial flat lands around Bordeaux today, the advantage of a fine wine over cereals must be of the order of many hundreds of times and even with ordinary *appellation contrôlée* much more than 20:1. But in the huge mechanized vineyards of the Hérault the ratio is probably about the same today as it was in Athens or Corinth, when both cities were Imperial powers trading with colonies all over the then-known world.

Greeks and Romans were said to fear the sea-journey through the Pillars of Hercules, known less poetically as the Straits of Gibraltar, but the Phoenicians went through and far beyond. It was they who took the culture of the vine to their colonies in Atlantic Spain, including Cadiz, and to Portugal. But the Phoenicians voyaged even further, and traded in Cornwall for tin, which was difficult to find in the Ancient World once alluvial tin had been generally exhausted.[4] As far as is known, they did not take vines to Cornwall, though they probably traded its product, wine. The Greeks took cuttings of the vine to their colonies in southern France, southern Italy and Sicily, and to the shores of the Black Sea. The benign climate of the Crimea became a favoured place for vineyards and wine for more than 2,000 years.

Each Greek city-state was individualistic and found cooperation with other Greek cities difficult, if not impossible, sometimes even when externally threatened. It is said today that if half-a-dozen Greeks were to be found together, there would be ten political parties. The Romans were very different. They did not believe in cooperation between equals, but once they had conquered another land, they encouraged a class of what we would now call 'collaborators' through whom they would then rule indirectly. They also settled retired soldiers in the new colonies on the boundaries of the Empire, as far as what is now Iraq, and North Africa, and on the Rhine, and in the North of Roman Britain, on Hadrian's Wall.

The Roman Imperial economy was intended to exploit the appropriate resources of conquered countries. The City of Rome had a population of over 1 million for at least 300 years and was a great consuming centre, rather as Manhattan is today. Here were the wealthiest and most powerful people in the world, and they demanded and obtained the best in wheat, olive oil and wine.

Egypt, Sicily and North Africa were required to grow wheat and barley for Rome; the vines of Carthage (now modern Tunisia) were destroyed, as were those of Sicily, except for a small area round Messina. Some old vineyards in the Roman colonies were replanted with olives, and vines were grown in suitable parts of Italy, Lombardy and Tuscany, much the same areas as those that produce the best wine today. Some parts of Spain, such as Andalusia, and suitable places in Syria were also planted with vines. Only later is there any mention of two great red-wine-producing areas of France – Bordeaux and Burgundy. This may be because the Romans seem to have preferred sweet white wines to robust reds. In the absence of sugar, and with the strong aftertaste when honey was used as a sweetener, people yearned for naturally sweet food and drink. If this theory is correct, the Romans wanted – and their biological systems needed or craved – sweet, syrupy wines, with high natural sugars, especially in winter when sweet fresh fruit was in short supply (no oranges or bananas were available in those days and ancient apples grown in a Mediterranean climate did not keep well in winter). Or maybe red wine did not find favour because keeping wine was difficult, and it was not allowed to age properly.

*

Wine and olive oil were kept and carried in *amphorae*, tall pottery jars with narrow necks for wine, wider mouths for oil, each of which held eleven Imperial gallons, or about 50 litres. In Rome, there is a huge mountain of broken *amphorae* called Monte Testaccio; this eminence is reckoned to hold the fragments of 40 million *amphorae* – though no one knows who counted them.

If the figure is correct, it stands testimony to the wastefulness of the great consuming city that was Rome. Unless *amphorae* are badly tainted, they can be used again – oil for oil, wine for wine – but in both cases the contents will be of lower quality than before. It is possible that the broken *amphorae* in the Monte Testaccio had been used several times and this was their final resting place. If used five times, these 40 million *amphorae* would have carried a total of 2 billion gallons, which sounds entirely plausible for a city of a million people using a dump for broken containers over several centuries. But today it is unknown at what date Romans started exploiting the Monte Testaccio as a dump and at what date it ceased to be used. What is interesting is the long-established use of *amphorae*, which were not the handiest of vessels. They had to be propped upright, often in special frames, since the bottom was pointed. Awkward to carry, arduous to sling on the back like a sack, they weighed over 135 lb when full, a fair weight to handle. Pouring was not as easy as from a jug or bowl, because the neck was not wide in relation to volume and the air did not conveniently replace the wine as the wine was poured. *Amphorae* were also difficult to glaze internally, so that wine might leak out and air might leak in – a fine recipe for turning wine into vinegar over a few weeks. Nor was a lid – even when sealed with a sort of canvas washer – a proper substitute for a cork, which was not always used, being expensive and difficult to obtain.

Why were barrels not in general use? They were made by Celts and other non-Latin 'Barbarians', but suitable wood for barrel staves may have been in short supply in Italy, or the skills of barrel-making may have been only known north of the Alps.

There is no real evidence in any of the rather sparse contemporary works about wine in Ancient Rome concerning the use of *amphorae*; nor is there much about vats for the actual making of wine. In more recent times vats have been of oak (or some other hardwood), cement, stone or masonry, more modern alternatives being stainless steel, fibreglass and metal tanks lined with glass. If the wine is to be good enough to keep six months beyond the spring after the harvest, one of the first essentials is to be able to cleanse the vats properly, removing yeasts and other micro-flora that have inhabited the air in the empty vessel since the previous season (always assuming it was properly cleaned after the preceding harvest). The Romans obviously could not use high-pressure steam for cleaning, but a vat's surface could have been of the same smooth cement found in the best Roman aqueducts and would have to be cleaned by slaves using brushes, sand or other scouring media and plenty of clean water, of which aqueducts supplied an abundance.

One must, however, sound a note of caution about this kind of speculation. In a parallel case, that of Roman furniture, many labour under the delusion that most of it was made of metal, not because this is a revealed truth but because the Barbarians who sacked Rome and all those who did not appreciate Roman artefacts burned wooden objects to keep warm, or used them to make something else out of wood. Even the civilized newcomers would often rubbish the antique as undesirable and 'old-fashioned'. Just as Roman buildings were deconstructed to make roads or to use as quarries for other new buildings, so wooden objects were recycled by succeeding generations to the extent that few Roman artefacts made of wood have survived. There is, in fact, more wooden furniture extant in Egypt, where it has been preserved in the very dry atmosphere of tombs, than in Rome, a metropolis many times larger than any other city of its time. But Rome was a living as well as an Eternal city. So wooden vats, if used, may have been subsequently broken up by Barbarians for reconstruction purposes, or the timber may have been burnt for warmth, for cooking, or even for blacksmithing or smelting ores. The example of the (now absent) furniture is a warning against too much speculation about the type of vats that were used.

*

After about AD 350, the Eastern Empire prospered; it was Greek-speaking and worshipped in what would become the Greek Orthodox ritual. Its capital was Byzantium, later Constantinople. Latin-speaking Rome, in contrast, used the Latin rite and became the home of the Roman Catholic Church. Politically and mili-tarily Rome went into grave decline and, after AD 400, was unable to defend itself. With much of the supporting hinterland overrun by Barbarians, the once great city sank not only in power but in population and most of the amenities of a sophisticated metropolis disappeared, including drinking water itself on several occasions. There was a struggle for survival, the nearest parallel in our own time being the state of Eastern Europe after 1945. Only the strong or lucky pulled through, as in East Berlin after the Russian victory in 1945.

Constantinople, in contrast, regarded itself as the true successor to Imperial Rome, strong enough to defy the Barbarians who ravaged the West. In the sixth century, Justinian and his General, Belisarius, defeated the Persians in the East – they had been opponents of Greece and Rome for more than 1,000 years. In the Mediterranean, Belisarius defeated the Vandals in North Africa, and the Ostrogoths in Italy. Later, Belisarius liberated Rome itself, causing such a degree of jealousy in his patron, Justinian, that it led to the General's dismissal. The disastrous dispute between Emperor and General was ruinous for the Empire and subsequently the West reverted to division and near-barbarism in places.

Although the people of the Byzantine Empire – Greeks, Levantines, Jews and Arabs – traded with the East, as far as the Malabar Coast of India for pepper, there was little trade in the Western Mediterranean. This area was about to be disrupted, in any case, by two dynamic, near-explosive waves of new strangers, Muslims from the south-east and Vikings from the north-west. Neither made wine. Muslims, within a century of the Prophet's death in AD 632, had moved out of Arabia along the North African coast, all over Spain and as far as Poitiers in Central France. Nearby, at a great battle in 732, the tide was turned by Charles Martel, and the Muslims retreated beyond the Pyrenees. They remained enemies of the Franks for another century, but Charlemagne, Charles Martel's grandson, united France and set up a Christian Empire (later to be the Holy Roman Empire). The Muslims were restricted to Spain and North Africa.

From the north-west came the Viking threat to Western Europe's efforts to regain the kind of order which had been lost when the Western Roman Empire collapsed. Vikings started negatively in the business of rape, pillage and piracy, but later became traders and settled in Ireland, Britain, Normandy, Portugal, Gibraltar and Sicily. They also colonized Iceland, Greenland and (it is said) Newfoundland. Others Vikings went via Nizhniy Novgorod (a city they built) and the rivers of Central Russia to the gates of Constantinople itself, more than a 2,000-mile trek from their homeland.

II

The making of even simple wine by simple peasants was dependent upon a degree of peace and calm that was rare in those troubled times. On the other hand, most of the conquering Barbarians found even the least sophisticated wine more to their taste than cruder forms of ethnic alcohol; in the north beer was made without hops which meant that it had to be brewed often as it did not keep well, and mead was made from honey. The Muslims did not, officially, drink any form of alcohol, stimulants having been banned by the Prophet, but the Vikings had a general genius for absorbing the culture of those they conquered, often because the conquered outnumbered them by several dozens to one. So wine-making continued in all the lands that had previously made wine before the arrival of the Vikings. They even called Newfoundland *Vinland*, or land of the vine, but only, presumably, in contrast to Greenland, since no modern has ever found it possible to make wine in Newfoundland.

All over Europe deals were struck between new masters and native peasants with wine-making skills, just as there had been many other deals made between military victors and skilled peasants who grew grain, made harness or were blacksmiths. The relationship may have started as the classic connection between master and slave, but it developed into the feudal system, best established in the West in France. Here, nearly everyone had a feudal lord, with five or six levels of lordship. The ultimate lord, often a king, called on dukes and earls to pay rent for land with feudal dues, providing the services of a fixed number of people as annual *feus* for occupying land. The provincial lord then demanded service of his own tenants-in-chief, and they in turn would claim *feus* of humbler tenants, and so on down to the the serfs at the bottom of the pyramid. Serfs could demand protection in exchange for labour for so many days a week or for the military services of the younger, more athletic and more skilled among them.

Serfs were attached to the land, not to masters as were slaves. Whilst a slave could be bought and sold like a horse or a cow, a serf could not be sold away from the land. When the land was sold or otherwise changed ownership, the serf went with the land to its new owner. A good master did not unduly exploit his serfs because, apart from any charitable concern, the long-term value of land depended upon the welfare of the serfs attached. Empty land without labour was of little or no value. So serfs were usually better treated than slaves and had the right to work their own land for so many days a week or so many weeks in the year.

Nevertheless, the lord's grain was usually sown first, the lord's cornfields tended to have fewer weeds, and the lord's harvest was more often than not gathered at an optimum time. The lord (or his sons) was said to exploit the *droit de seigneur*, an imagined right to impregnate any woman within his feudal curtilage.[5] But there was a much more important element in the relationship of serf to lord. A lord might well regard serfs as children, even if they were biologically unrelated. This is exemplified within living memory within some Highland clans. The laird, if living amongst his own, might regard the welfare, education and advancement of his clansmen as his responsibility as much as if they had been his biological children. The same sometimes applied in feudal times, even if it meant that lords (and their villages) lost the services of the cleverer young male serfs, who might go off to be trained as priests, the only people in the Middle Ages guaranteed to be literate.

To train for the priesthood was, of course, a prime means of worldly advancement for the brighter sons of every class (even of the highest noble classes) in the early Middle Ages, the only way to guarantee an education being to join the Church. Priests in the West were meant to be celibate and to die without heirs or any material wealth to pass on when they left this world. The original philosophic reasons for the vow of celibacy were to prevent any tendency for clerics to be worldly. So it was a curious fact that all those who could read and write (nearly all clerics) were unable *legally* to pass on their genes to another generation. Put another way, the genes of those able to read and write were snuffed out in every generation without descendants.[6]

Serfs bound to manors, abbeys and monasteries laboured in the vineyards attached to the great houses: at first these were wholly clerical; only later were there lay vineyards. The circumstance that the Christian Church held that bread and wine were a central feature of the Faith is essential to the history of both wheat and wine. The most important Roman Catholic ceremony, the Mass, made Western Europeans the great vintners of the world, even if some of the best wine today is made by non-clerical, post-Christian people in the Neo-Europes of California, Chile, Australia and New Zealand.

There were and are, of course, monastic foundations in other religions – there are Buddhist, Hindu, Jewish and Muslim monasteries. But as from the foundation of the great Christian monasteries, starting with Monte Cassino, between Rome and Naples, in 529, just over a century after the Fall of Rome in 410, monks became the conservators, practitioners and teachers of the skills of husbandry. In a nearly entirely agricultural age, monasteries held the essential key to human survival, a sufficiency of food to eat, second only in importance to an absence of threat to life and limb.

It is arguable that without the monasteries, the art of husbandry of fields, gardens, orchards and vineyards would have been lost in Western Europe. None of the new militarily successful but savage tribes that displaced Romans and Romanized natives had any of the necessary skills or expertise to make wine, let alone good wine.[7] Monks carried on the practice of selecting, growing, picking and crushing grapes to make wine. Far more than laymen, they held the whole future of wine (as well as of all other forms of agricultural husbandry) in their hands. For more than 500 years after St Benedict founded Monte Cassino, it was monks who transmitted the arts that Greeks and Romans had developed to grow grapes and to make wine. After AD 1000, however, some of the laity commercialized wine. By the twelfth century the Bordeaux region, for example, was full of lay growers, makers and exporters of wine, and an important seaborne trade began with Bristol and London.

The ships were usually manned by native Gascons and they sailed as far north as Antwerp, beyond which port the Hansa merchants ruled the waves of the northern North Sea and the Baltic. Since the Millennium Year of 1000, trade had accelerated through Europe, as a few wise men recognized that trade could bring far greater wealth than booty won in war. The Alpine routes enabled Burgundian or Rhenish wines to be sold in Venice and to be exchanged for pepper, which had arrived by pack-animal, ship and mule-train from Southern India. Wool and cloth from England and Flanders were one part of the trade trio of wine, salt and wool, and it was only wool and wool-cloth exports that allowed the English to import as much French wine as they did. Even though there were more than forty vineyards listed in England in the Domesday Book of 1087, the Norman tax-census of the land they had conquered, no one today knows anything of the quality of the English Domesday wines, but though the climate was probably warmer then, it is likely that most northern vineyards only produced sacramental wines for Church use.

*

At some time before 1300, the Atlantic weather started to cool. Wheat was no longer grown in Greenland; wine was no longer made in England. It is now clear that at the same time, Europe had reached a peak of population that brought critical food shortages within the agricultural husbandry of the day. After some harvests, there was widespread hunger; after others, famine. Often local, famines usually afflicted those districts without the wealth or the skill or transport to make good any deficiencies.

Then came the Black Death, which killed perhaps one-third of the population. Population numbers did not regain their 1348 level for nearly 150 years. The positive side for survivors of the plague was that 'hungry' and marginal land had no longer to be cultivated with back-breaking effort, never usually justified by the meagre results. Every village had nearly half as much more land per head to produce wheat, wool, meat, oil or wine. Of all these products, wine was the most valuable, creating wealth at every stage from vineyard to consumer.

From the 1150s until the 1440s, England controlled most of the wine-producing areas of western France. The English then drank as much as a gallon of wine per head per year, more than

the national average today after centuries of change and many
boom years, with wines imported from all over the world, not only
from Bordeaux and Touraine. After the Bordelais became French
again, the English started to import Burgundy via the port of
Rouen and, soon after the Renaissance, wine from other European
countries. The Renaissance and the Age of Discovery coincided,
and not the least coincidental side-effect of locating the sea-route
to India and finding the Continent of America was to make
Europeans think of alternative ways of doing things.

III

By about 1500, the great vineyards of France and Germany had
assumed a pattern that would survive until after the Second World
War. Professional wine-making depends on three stages: first, to
grow the grapes; second, to ferment their juices; third, to save the
fermented product (nowadays in glass bottles) and to care for the
contents until it reaches the consumer. In each stage, opportunities
for error are very great, but the best husbandry has been ordained
for centuries. An adherence to good practice was (and is) much
more profitable than expensive efforts to correct mistakes.

The first stage, the growing of the grapes, was in some ways
beyond the usual options. Although vintners could choose the
grape variety, follow the dictates of good management and avoid
disease, little could be done until recently about the nature of the
soil; even now, less can be done about climate and weather.

Wine-grapes are grown today in Europe as far north as Bonn
on the Rhine, where Chancellor Adenauer was the owner of one of
the most northerly vineyards. The latitude of his vineyard was
parallel to Sussex or Hampshire, about 50° 50' north. The southern
and western limit for growing grapes is in Atlantic Morocco, where
not very good wine is made. The total European range is thus
about 1,500 miles. Similar limits exist in the East, from the Black
Sea to southern Israel, where the climate is, as in Morocco, really
too hot for good wine.

From 1400 onwards, vines followed the explorers − to the

Eastern and Western Atlantic islands, and later to the American
mainland. In colder areas, grape-juice tends to lack sugar and to be
too acid; in excessively warm areas, the product is flaccid and
prosaic, but can be made into a good fortified wine like Port,
Madeira or Sherry. The best European wines come from Bordeaux
and Burgundy in France, near the Rhine and Moselle in Germany
and a few places in Italy, where the climate is neither too hot nor
too cold. But everywhere the weather, as opposed to the climate,
plays a major role. In every vineyard, a warm spring is needed to
form plenty of flowers and to set the grapes, while a mixture of sun
and rain is essential to swell them and build up the sugar-content.
Late spring frosts are of course unwelcome and can be damaging,
reducing yields and sugar-content, and in the worst scenario
inducing a condition known as acid rot, which cannot be reversed
by later sunny weather or by human intervention.

Sweet white wines thrive in hot summers better than do reds.
Whilst a high-sunshine summer will produce a great Johannisberger
or a great Château Yquem, the same weather will give red wines
made from grapes grown next door a 'burnt' taste. Contrariwise,
cool, wet summers will produce poor, thin, acid wines, short of
sugar and without the indefinable flavours that make a wine
worthwhile. Weather, as opposed to climate, is often very local:
one side of a hill may miss a hailstorm; and another vineyard,
a few hundred metres away, will benefit from a shower at a
crucial moment in a hot summer, so that the grapes are refreshed
at a critical time. Without that shower the grapes would have
shrivelled, as perhaps did those in a less fortunate neighbouring
vineyard.

Topsoil is less important than might be thought. Champagne
grapes are grown on thin soil above chalk; in the Bordelais, the soil
is alluvial gravel over clay or sand, or a mixture; in the slopes on
the sides of the Moselle Valley, the topsoil is largely slate. What is
important in all these places is that the soil should be friable
enough for the vine's roots to penetrate, since it is deep roots
tapping the micro-nutrients that give good wine its character and
there is, of course, usually more moisture deep down. Near-solid
rock is unsuitable as a subsoil for vines, as it is for virtually any
other crop, especially in hot, dry climates.

Vines are planted in straight lines more than a yard apart, with wider gangways favoured if greater mechanization is employed; in Australia rows of vines are often as much as 3–4 yards apart. Different methods of training vines are employed in different areas, the unrestrained nature of vines being to sprawl and climb wherever it is possible to go. The aim is to limit woody growth and to prune in the early spring to two spurs, one to develop a bunch of grapes for the same year's harvest, the other to produce wood for the following season.

Some form of weed-control is needed between rows of vines and this may take the form of ploughing or cultivating, which is often preferred to chemical control. There may be two, three or four cultivations of the soil between the rows of vines. Grapes grown far from the soil (too high up) tend to be lower in sugar and contain more acid and moisture than is desirable. This is because the soil reflects the sun's heat and light and at night retains the heat of the day and keeps the grapes warm. Sugar-content can be very local, both in time and place: grapes five feet above ground are lower in sugar than those close to the soil. A really cold shower during the harvest can reduce sugar-content by a third. A delay then becomes essential so that grapes can recapture the necessary higher sugar-content before they are picked. Conversely, in very hot seasons, in southern vineyards, there will be too much sugar and too little acid in the ripe grapes, so a clever, cheap device is employed to restore the balance between sugar and acidity. A few unripe grapes are routinely picked early, crushed and the juice preserved for a few days or weeks so that the grape-must has enough acid to make a good wine, the acid/sugar balance being essential to any good wine.

Subject to weather, grapes are usually picked between mid-September and mid-October in the Northern Hemisphere; there used to be a fixed date for the start of the harvest in most districts. A Greco-Roman rule forbade any grapes to be picked before the rise of the great star Arcturus, which happens every year at the end of August. This day became an important festival, including song and dance, with work starting the following day. The date was enshrined in law in the Middle Ages in ex-Roman wine-producing areas and in some places, notably in Bordeaux, the rule lasted until

1889. In other regions of France, and in Italy and in Germany, local authorities were given the right to set the date of the grape-harvest, which was then religiously (often literally) respected. Nowadays, rules like this are still honoured in most areas and, following a cool, sunless season, it has been known in Germany for harvests to be as late as November to catch every ray of the autumn sun. In at least three seasons since 1950, in a few German vineyards, a few grapes have actually been picked as late as New Year's Eve. This has had nothing to do with the making of the very sweet white wine, however, when ripe grapes are left on the vine to develop *botrytis cinerea*, a mould that produces 'a noble rot', or *la pourriture noble* or *Edelfäule*. The mould reduces the moisture content of the grapes so that sugar-content rises to as much as 60 per cent from a normal 20–25 per cent. These grapes produce wines that are no more alcoholic than normal but much sweeter, because the enzymes in true wine yeast cannot convert more than about 25–35 per cent of sugar to alcohol. This leaves the remaining sugar (up to 60 per cent) to sweeten the resulting wines greatly. With temperatures as low as −5°C when the grapes are picked, the wine is even more valuable and is called *Eiswein*.

The *vendange* or grape-harvest is traditionally a time of good feeling, joy and laughter, the work in the sun being congenial, and a special effort is made to feed the pickers well and to allot them sufficient wine to help the work progress but not enough to encourage anyone to doze off.

Grapes are hand-picked in many great vineyards by experienced people using traditional knives, curved for convenience, or the more modern secateurs, tools invented for this very purpose – which have now migrated out of vineyards into gardens worldwide. Traditionally the pickers were often women, while men were responsible for carrying grapes to a wagon, ox-drawn for preference. If men were stronger, women were more careful about selecting the fitter, riper bunches of grapes, while oxen were more patient (especially where there were troublesome flies about) than horses or mules.

On the very steep hillsides of some of the West German, Swiss and Italian vineyards, pack-horses or pack-mules were used even in recent times to bring harvested grapes to a road where they could be transferred to a wagon. In the first peace-time Rheingau harvest in 1946, pack-animals, their ancient Mediterranean origin at least 3,000 years old, carried precious grapes to surplus US Army trucks. These trucks had originally brought benign invaders to rescue Germany from the Nazis, as one post-war story had it. Few in that Rheingau vineyard would have described the last bloody year of the Second World War in quite that way, but that first post-war harvest was a time of hope after the nightmare of the Hitler years.

IV

Consuming wine was associated, since early historic times, with well-being, *bien-être* or *gemütlichkeit,* and was an inspiration for song and verse. Wine was offered to guests and travellers, to be drunk alone or with water before, during or after meals, as a symbol of peace and friendship. But wine has been more than all that.

It was an early acceptable form of mild narcotic, a gentle soporific, inducing euphoria, a liquid full of sunshine, a gateway to happiness, a unique friend softening a hard life. More than all this poesy, fermented wine or beer was often the only safe drink, being free from parasites. The vine and its chief product, wine, very early indicated four economic roles in the Mediterranean civilizations, and never lost them.

First, wine symbolized stable husbandry. Vines are medium-term investments, not as far-sighted as olive trees, but much more is implied by the planting of vines than the clearance of land for cereal production. A vineyard has an economic horizon of at least forty years, as much as a lifetime in ancient times. There was a wait of five years before full production. This meant that an adult man planted a vineyard for his successors, not necessarily for himself, and the sense of stability this involved made vines and wine an essential indicator of a certain level of civilization, of a certain belief in security, of a certain degree of faith in the future.

Second, wine was traded from the very earliest times. It was traded for gold, slaves, many times its weight or volume in other agricultural produce. Thus it was an object of value for trade, and sometimes a store of value. It was traded as early as 500 BC over great distances: from Greece throughout the Levant; from Egypt down to the Horn of Africa and perhaps beyond, as far as Zanzibar; from Persia to the Bombay coast of India; from Phoenicia to the Western Mediterranean; from the Greek colonies in what is now modern France, all over the Middle Sea; throughout the Roman Empire. Trade in wine is an indicator of something of great importance, for whenever there is evidence of a wine trade in history, there is evidence also of a degree of peace and economic order. To make wine of export quality was not the only problem. Almost anyone could make wine, and did so, if better, imported wines were not available. The northern wine–vine limits are not very different from the apple limits and wine was a more acceptable alternative to cider in the higher civilizations, although reasonable cider could be made where wine would be too acid. Imported wine implied a sophisticated taste, and a trade in wine indicated at least three factors: a buyer who could be persuaded that imported wine was superior to local drink; a trader who felt confident enough to travel with easily consumed or destroyed merchandise, therefore trade-routes of a reasonable level of security; finally, and perhaps most important, a vintner or merchant in the wine-producing area who could sophisticate the wine sufficiently, in the days before sugar, to allow it to travel long distances. Evidence of a considerable wine trade implies the creation of wealth

Third, and very shortly after the second stage, when a regular trade in wine was established over the routes already described, sophistication of every type set in. Until recently, most 'peasant' wines had to be drunk in the year following production. By May–June of the year after a Northern Hemisphere harvest, most ancient wines would be turning sour. In order to carry out a generalized wine trade, crude chemicals had to be used as preservatives, without knowledge or reference to customers' health or safety. Lead, antimony and sulphur were in use before the Christian era, probably before 300 BC.

For a people whose city's inhabitants were being slowly poisoned by the lead in drinking-water pipes, the lead in the alternative to water – wine – may not have worried the Romans overmuch. Good 'pure' wine could have been made in Greek colonies in France, Italy, and Sicily before the Roman take-over, and good resinated wine, the retsina of today, was being traded from Attica itself from about 300 BC, and was an important element in the Athenian balance of payments. No one knows much about the proportion of the adulterants, or the level of sophistication of the wine itself, but no modern wine-buff, whether vineyard owner, merchant or consumer, can imagine what the experience of a draught of wine drunk far from home can have seemed like to the ultimate consumer in the Ancient World. To be fair, water was probably always more contaminated than any wine unless the water drinker knew his source. It was the pollution of nearly all water which made it unsafe, in an era before tea, coffee or other hot drinks. So wine was an essential for all but the natives used to the polluted water.

Sophistication went further than the adulteration of wine. Trade in wine has always involved its 'improvement' by means other than the mere addition of improving agents. Although pure, unadulterated wines were already hard to find in Virgil's day, and comparisons with virgin girls were made by both Ovid and Horace, the language used to sell wine was already an extravagant joke several hundred years before in Athens. Language describing wine has rarely been as pure as wine itself and has never been in the least bit restrained. The wine-speak of today is nothing new, and has a respectable history of at least 2,500 years. There is a magic imputed to wine, even today, when mankind can seek alcohol in many different forms; wine is still considered the more civilized, more up-market alcoholic drink. To assist that image, the trade has enlisted culture and poetry; chroniclers, heralds and bards have been employed to lend a hand to the merchant's efforts. There is nothing new about the wine snob using wine-speak. Nothing today can make him any less funny than he was when the Ancients laughed at him, engaged in the same exercise, all those years ago.

Fourth, and finally, mankind could choose wine as an alternative crop to grain in the very early days of ancient Greece. In adopting the change-over, Greek cities were making a deliberate choice; they were opting for what was then a market economy. It was an event of great importance, the first evidence of the specialization, the division of labour, the performance of the task best suited to the circumstances and to the talent of the manager, which Adam Smith was later to applaud. The deliberate change-over by some Greeks from grain to vines has an almost modern look. Yet the decision was made in the time of Pericles in ancient Athens, and from this choice flowed massive consequences.

Owners of vineyards became wholly dependent upon trade, and it had to be successful trade, which would in turn imply colonies growing grain which would be paid for by wine. There were over a dozen Greek colonies in the Black Sea by 400 BC, nearly a hundred in the Mediterranean two centuries later. The pattern was that wine would be grown in the poor Attic soils many times as profitably as grain grown in the same soil, and would be exported in merchant ships, protected from pirates, all over the then-known world. In essence, it was a very simple form of national 'gearing'. An area in wheat, in the thin Attican soil, would produce less than 500 kilograms of wheat to the hectare. This should be compared with the possible 2,000 kilograms per hectare grown by the methods of the pre-Christian era in the Crimea, or on the Danube Delta, in the alluvial soils of the Po Valley or of the Pas de Calais, by methods we know were employed by the British Celts. The soils of Attica, worthless or virtually worthless in grain, were 'priceless' in wine, probably growing, in terms of value, the equivalent of 20 tonnes of wheat per hectare. So to grow wine was to increase the potential value of the product of the land forty times.

It was not only wheat that wine bought. It was traded, as already stated, for slaves, which were not only a source of labour but also a form of currency, a store of wealth, objects of prestige. Wine was also exchanged, of course, for jewels, gold and other metals. It became a truism that the remote, scarred hillsides, relatively profitless in grain, when planted to vines became more valuable than the richest, strongest, finest alluvial soils in the world.

This is still true today. In some areas, the poorer the land, in conventional chemical and physical terms, the more valuable it is if located in the right area topographically. With the right aspect, an appropriate climate, seasonal weather, rainfall and sunshine, poor hilly land is worth far more than flat grain fields. For hundreds of years, land which in most other crops would ensure bankruptcy has guaranteed a competence, prosperity, even wealth, when in vines. This is true nowadays of large areas in Bordeaux, Burgundy, Alsace, the Rhine, the Moselle, and the Tuscan hills. As is said today, in some areas of Burgundy: 'If our land were not the richest in the world (for wine) it would be the poorest in the world (for anything else).'

V

Grapes are one commodity, wine is another. In a truly natural fermentation, without the intervention of man, a recycling rather than a mere fermentation takes place. Grapes, as picked ripe in the field, contain sugar, formed in the process of photosynthesis through the summer from carbon dioxide and water in the leaf of the vine. The grape skins are coated with millions of yeasts, bacteria and moulds. These invisible (and even now not wholly identified) basic forms of life exist in all the 'pure' air which we breathe, and in every part of the countryside, anywhere in the world. They carry the appropriate organisms to encourage metamorphosis in the living world. They are the precursors of growth, decay and change. Yeasts are beyond number – an actual count of 1 kilogram of grapes in a mature vineyard in California totalled about 100,000 moulds, 100,000-plus 'wine-yeasts', and more than 10,000,000 'other' yeasts.

If grapes are crushed and left to ferment without interference, the wild wine-yeast (*Saccharomycea apiculatus*) will take charge, and convert some of the sugar into about 4 per cent alcohol. This concentration of alcohol will then kill the yeast, which has done its work.

The wild wine-yeast is superseded by the true wine-yeast (*S. ellipsoideus*) which takes over and continues to convert sugar to alcohol. The take-over does not occur suddenly at 4 per cent alcohol, and there is considerable overlap, but, by the time *S. apiculatus* is dead, *S. ellipsoideus* has begun working with great vigour.

Temperature is critical. At work the yeasts give off carbon dioxide and heat, just like a human athlete. The wine should be kept below 25°C or the yeast will be killed, and refrigeration is employed in hot climates. In October in areas like the Rhineland or Burgundy, the autumn weather may be so cold as to endanger the yeasts, with temperatures falling below 5°C, at which point yeasts hibernate. These are extremes.

Within this 5–25°C range, a high temperature produces a quick fermentation which endangers the wine-yeast when exposed to air. A slow fermentation results in much more subtle flavours but carries the risk of allowing the dread acetobacter to start its deadly work.

Acetobacter is an organism which turns the alcohol in wine into acetic acid (raw vinegar). This will not normally work contemporaneously with the wine-yeast, but if the process remains unchecked, the acetobacter will take over at some point and turn all the alcohol produced by *S. apiculatus* and *S. ellipsoideus* into acetic acid. When that process is complete, the acetobacter is destroyed by the vinegar. The neatness and elegance of this biochemistry fascinated people long before they understood it. The wild yeast produces enough alcohol to kill itself, then the true wine-yeast turns all the sugar into alcohol, up to a concentration of about 12–14 per cent, which then kills the true wine yeast. Then the acetobacter takes over and converts alcohol to vinegar, which in turn kills the acetobacter. The vinegar (acetic acid) is not stable. Left to itself, it is attacked by the true nasties, moulds and bacteria which putrefy and do not ferment. These gradually break down the acetic acid into CO_2 and H_2O (carbon dioxide and water), which is what the vine's leaves and grapes contained, and which is where the process began. Each process in turn contains the seed of its own destruction, as Marx (wrongly) said of capitalism.

The process of growth from leaf to grape takes six months or so. If a bunch of grapes were left on the soil under the vine in weather warm enough to permit biochemistry, the process of decay would take from two weeks in very warm weather to two months in a cool autumn. Thus does Nature ensure cycling and recycling, and it is characteristic of all the processes of growth and rot that the rot is much more rapid than the growth. If the reverse were true, there would be very little growth. A moment's thought will prove why. Vegetable growth draws energy from the sun and water from the soil, but also needs minerals and trace elements in readily available quantities, however small, and these are obtained from humus, the top layers of the soil, mixed with rotting organic matter. So it has to be that the rot (or recycling) is quicker than growth.

After picking, the wine-grapes must be crushed, and the same process has been followed for 2,000 years or more.

If red wine is wanted, the grapes will have been crushed 'on the skin' to obtain the colour, tannin and taste, and the solid mass of skins, pips and sometimes stalks is held up in the top of the vat by the generation of carbon dioxide which is one by-product of fermentation. This cap of organic matter must be constantly immersed to prevent an increase of acetic acid. (Already the wine-maker is interfering with nature.) In huge modern wineries, the stalks are not included in the press so that the mass can be pumped round and round and kept in solution. The stalks are not included in the better red wines of Burgundy or Bordeaux, nor are pumps employed because, in the production of fine wine, stalks and pumping would lead to an uncontrolled release of tannin from stalks, skins and pips. So the cap is either trodden or pushed manually with poles, or a mechanical analogue of treading is used. Manual treading is very hard work. The cap is so solid that a man could be supported in a vat of anything more than 10,000 litres but, in order that he should not die a fragrant death by drowning, or by carbon dioxide inhalation, a bar is fixed for him to hold. Even so, intoxication used to be common, either from the fumes or from the CO_2.

No Health and Safety inspector today would allow the process of manually treading down the cap. However, although treading the cap may be very hard work and occasionally dangerous, it is unexcelled for cleaning the feet and destroying the fungus which causes athlete's foot, just as hand-washing of clothes ensures clean hands. (Natural processes are full of beneficial side-effects.)

But there are problems. The first is how to get the wine up to 4 per cent alcohol as quickly as possible in order to ensure that it enjoys true fermentation at 4–12 per cent alcohol and does not suffer from any bitterness left by the wild yeasts. As with many other familiar processes, this chemistry only became formally evident just over a century ago. The acceleration can be achieved by heating the vat to 18–23°C (the optimum fermentation temperature), thus cutting down the dangerous period when correct fermentation is at risk, at below 4 per cent alcohol. There is an alternative, an ingenious method used in Italy, but invented by a Frenchman, Semichon. That is to add one part made wine of 12 per cent alcohol to two parts of the must, which contains no alcohol, so that the mixture starts as a liquid of 4 per cent alcoholic strength. The trouble with the Semichon method is that any biological fault in the added wine is spread into all new production. So the process is not normally used in making fine wines.[8]

A third, purely chemical, method was at one time widely used and abused. This involves adding one part of sulphur dioxide to 10,000 parts of the must so that the wild yeasts are all killed, along with most of the moulds and bacteria. True wine-yeasts (*S. ellipsoideus*) are then added. This is easy and sure, and relatively cheap, but wine-drinkers the world over claim that sulphur can be tasted, even in tiny concentrations. A further claim is that if 100 parts of sulphur dioxide per 1,000,000 (one part per 10,000) are added to the must, then more than this proportion can be left in the wine in the bottle. Anything in the region of 250 parts per million can be tasted by a tyro, and an expert can taste 50 parts per million – so it is thought to be far better to do the job properly. But some wines do taste of sulphur, just as some cheap convenience food tastes of sodium matabisulphite, and for the same reason.

During fermentation, it is possible to exclude the air and prevent production of the unwanted acetic acid and some of the moulds and bacteria, by keeping the vats covered and allowing the CO_2 to escape through a one-way valve – this was, traditionally, a vine-leaf held down by a stone. Unfortunately, this method also kills some of the unknown micro-flora which give rise to the most delicate of the many flavours in fine wine.

It was discovered in 1897 that it was not the yeasts which did the work, but the substances within them, simple proteins called enzymes. Enzymes are very small. A yeast cell $1/200$ millimetre or $1/5,000$ inch across will contain in turn 1,000 different enzymes. Enzymes are natural catalysts, that is, they cause change without suffering any change themselves. They are very powerful. One drop of rennet, the extract from the stomach lining of a calf, can curdle a million drops of milk in an hour. Enzymes convert starch to sugar in plants and man; they convert sugar to alcohol; they convert alcohol to vinegar.

Until the great Louis Pasteur turned his attention to the problems of fermentation, no one knew anything scientific about yeast or enzymes or how fermentation occurred. Pasteur started the work of discovery in 1861, on wine. He arrived at the conclusion that yeast assimilates oxygen in order to live (correct). From this was derived the vacuum-packing of food, to deny oxygen to the agents of change. He also arrived at the conclusion that yeasts will individually only work on a narrow temperature range (correct). The pasteurization of milk makes it easier to bottle, distribute and keep, but the milk industry owes this elegant and commercial convenience to Pasteur's work on wine. It is no coincidence that the heat selected for the pasteurization of milk is the same as that which kills harmful bacteria in wine.

Pasteur also believed that yeasts (he did not know about enzymes) operated in wine by consuming air, so that, if you exclude air you prevent any deterioration of the wine (now obviously correct). But it is not only air that the yeast or the enzyme uses. It is also that amount of sugar which is not converted to alcohol. In the wine industry the *type* of enzyme has become of great recent importance. It is now known that an enzyme, about one-billionth of a millimetre across and invisible to all but a few microscopes, will make all the difference between a good wine, a fine wine and a great wine.

Work on the fermentation of wine unravelled the mysteries of yeasts (which in turn contained the enigma of enzymes) and produced some knowledge of vitamins, which are similar to enzymes. The range of esters, the function of phosphoric acid in life and, above all, the limits of economic hygiene in the food industry were all uncovered by the work of Pasteur and his followers in wine. The debt which we all owe wine, whether we drink it or not, is immense.

*

The process of fermentation is such that the grape must in the vat will contain, say, 24 per cent of glucose/fructose, and this 24 per cent will become (say) 12 per cent of CO_2 (carbon dioxide) and

10.5 per cent of alcohol. The balance (1.5 per cent) is lost in the process, primarily because the yeasts use sugar for energy, but also because they convert some of the must into glycerine, as little as 0.5 per cent in an early-picked crop of grapes but as much as 3 per cent in a crop gathered late.

If each vat of wine-must contained only 10.5 per cent of alcohol, as in the example quoted above, few would buy it. Industrial alcohol would be much cheaper, but within the liquid are live and dead moulds, bacteria and yeasts, and pectins, albumen, acids, glycerine, tannin, colouring matter, unfermented sugar and various salts of which the most obvious are the phosphates. These items may account for less than 1 per cent of the raw wine, but they are the only justification for taking all the trouble which attends the vintage, although the customer may still be years away from enjoying the efforts of the wine-maker.

When the primary fermentation in the vat is over, the wine may be filtered and pasteurized (or sulphured) before it is sold. The vast majority of wines used to be consumed in the year following the autumn of the vintage. Some of the higher-quality wines of immediate consumption are still alive when the cask is drawn, but most *vin ordinaire, Konsumwein* or *consumo* is dead when drunk. Fine and great wines, now less than 5 per cent of all the wine made in the world, are different.

The ability to improve in bottle is the deciding factor which makes a fine or great wine. Few white wines improve at all, and many deteriorate. Red wines of low quality also decline: of the middle-range reds, most stand still; few improve. It is only red wines of quality which really mature well in bottles.

The wine must be run off from the vat in which it has fermented and be separated both from the sludge at the bottom of the vat and the cap of spent must on the surface. The wine will still contain live yeast, and perhaps 1 per cent of sugar. The yeast will turn the last of the sugar into alcohol, unless a sweet wine is preferred, in which case fermentation has to be halted at the desired point of sweetness.

Fine or great red wines are run into barrels which line the cellar walls and weep gently at the seams. This weeping, which is more serious with the dry casks at the beginning of the season, has to be made good with wine of the same year or of the year before. Ullage or *ouillage*, the replacement of lost wine, is done daily at first, then every few days, and then becomes necessary only once a month or so. Various devices, well known to the home wine-maker, can perform this task automatically.

Solids in the wine gradually settle on to the bottom of the cask, forming a new sediment, and the wine must be 'racked' from this sediment. Racking involves drawing the liquid out of the barrel from above the sediment and separating it physically and chemically from these solids.

In much of Europe this is done three or four times in the wine's first year. In hotter countries racking is more frequent, because heat speeds up chemical reactions. In Africa, Australia or the Americas, it may take place as often as once a month. There is a great deal of learned discussion about the timing, frequency and severity of racking.

In warm countries in the Northern Hemisphere, racking needs to be done in October, both after the vintage and again almost every month. In Bordeaux, racking used to be recommended once during the winter and then in March and June or July, and per-haps in the following September. Whatever schedule is chosen, early racking cannot harm a wine, but racking too late can ruin a wine and cause a loss – financial, moral and, above all, a huge loss of face.

As well as racking, good wine needs 'fining'. This process encourages particles of yeast and other organisms too small to be filtered and which in the ordinary course of time do not form particles large enough to settle, to drop to the bottom. Or this is what appears to happen. What actually happens, as was only recently discovered, is that the particles of the fining agent have negatively charged ions, while the floating and minute debris is made up, as is most dust and detritus, of positively charged ions. The negatively charged ions attract the positive, and then combine to become heavy enough to fall to the bottom of the cask.

Some fining agents are 2,500 years old. As usual, the choice depends on the trade-off between cost and quality. Fining agents nowadays include bentonite mud-dust (borrowed from the oil industry), casein, milk, certain pure clays, white of egg, charcoal, blood (including the famous ox-blood), fish glue, gelatine and the seaweed derivative, alginate. Egg white beaten up with salt is the most expensive and the most favoured for fine reds. For white wines, isinglass (purified fish glue) is the correct natural substance – never glycerine, of which there is often too much in the wine already.

There are innumerable trade mixtures and compounds now available for fining, including many synthetics. It is theoretically possible to carry out fining by generating negatively charged ions electronically, thus purifying the liquid, as air is electronically purified, by making the particles heavy enough to stop floating.

Racked several times, fined at the proper time of year, in the proper weather (a thunderstorm during fining will scatter positively charged ions throughout the district and delay the fining process) and given time for the last of the lees to settle, the wine has only three major bio-chemical problems left. These are *disease, acidity* and *alcohol.*

Disease can be carried over from the vineyard, or introduced by lack of cleanliness or care in the pressing or racking or fining. In most cases, there is nothing to be done but early pasteurization and a quick sale under some other label. Yet disease is rare if every effort is made to avoid trouble. The key to success, as always, is dependent upon performing each operation in sequence with all due care, concern and attention to detail.

In every traditional establishment, however carefully the disease problem has been avoided, the balance between alcohol and acid is crucial. It is a difficult and touchy problem in every vintage, but briefly, as sunshine produces sugar in the grape, the more sunshine, the more sugar, and the more sugar in the grape at picking, the more potential alcohol there is in the wine. Alcohol is needed for preservation of the valuable micro-elements which give wine its taste, and a practical minimum is of the order of 9–10 per cent. Wines with less than 9 per cent alcohol will rarely travel, let alone keep. Wines of much more than 14 per cent have an alcoholic strength which makes the wine less than subtle. Inadequate alcohol in a vintage following a cool, sunless summer is the major problem of marginal vineyards, and is usually solved by chaptalization – that is, by adding sugar before or during fermentation. The practice is not widely talked about, nor does it attract approval, but if properly carried out, and sufficient time allowed for the enzymes to convert sugar to alcohol, there can be little chemical complaint. It is, however, considered a weakness to have to sugar the must, akin to making honey by feeding sugar to bees after a sunless summer.

Some people complain of wine being 'acid'. Yet without acids there would be no flavours in wine beyond the one or two present in grape-juice. The balance between alcohol and acid is a key factor in wine production and taste.

The vinegar taste can be prevented by correct fermentation, by preventing the must from being attacked by acetobacter, and by preventing the bottled wine from coming into contact with aceto-bacter. This means sterilized bottles and sound corks. Acetobacter is present everywhere, ready to turn wine, or cider, beer or fruit juice, into vinegar. Lament is not in order, since all forms of life should be allowed to live, and most forms live by change.

The other acid tastes are tart, not vinegary, and occur in wine because of excessive tartaric or malic acid. Tartaric acid is critical. Without two parts per 1,000 (0.2 per cent) wine will not keep. With more than eight parts per 1,000 (0.8 per cent) the taste becomes obvious, and above this figure wine is often almost unsaleable. Excessive tartaric acid is due to insufficient sunshine the previous summer. During the growth and ripening of the grape, acids form part of the process which yields sugar, and are ultimately (but not wholly) replaced by sugar. The boy who eats unripe fruit is ingesting tartaric acid voluntarily, but the vintner may have to pick his grapes at an imperfect stage because of inadequate total hours of sunshine. Since excessive tartaric acid is a function of insufficient sunshine, weather and latitude are important. To a lesser extent, unsuitable grape varieties produce excessive tartaric acid but choice of grape can be much more positive than the accident of weather. The problem of excessive acid is, therefore, largely a problem for northern vineyards and elsewhere after sun-deficient summers. The amount of tartaric acid in the wine as it runs off from a barrel is easily measured, and the correct residual figure may be obtained by care and husbandry, without the addition of any chemicals. The process is more than 2,000 years old.

Tartaric acid in wine can combine with potassium to form potassium bitartrate or cream of tartar. The amount dissolved in wine is proportional to the temperature. At low temperatures the cream of tartar breaks down, and the tartar forms crystals which fall to the bottom of the barrel or bottle. Ordinary racking in the cool of the cellar in the winter following the vintage, if properly carried out, will remove much of the tartar. Thus is virtue rewarded.

In hot climates, wine may be cooled by refrigeration, or even by cold spring water run round the barrels. In Ancient Rome wine flasks (*amphorae*) were put into streams (or snowdrifts in winter) for the same purpose. Whatever method used, there is little excuse for tartaric acid beyond that point needed for preservation. It should be borne in mind, though, that the more tartaric acid, the longer the life of a red wine and the greater the opportunity for such a wine to improve in bottle with age.

The other important acid which tastes unpleasant in excess is more interesting. This is malic acid, the acid of apples, which is present in grapes before being converted by sunshine into sugar. Malic acid is a food of the organism which converts CO_2 and sunshine into sugar. In a sunless year, or in a northern climate, not all the malic acid is converted into sugar before the grapes are picked. Provided that neither acetic nor tartaric acid is present in excessive quantities, a further wonderful transformation takes place. Malic acid is converted into lactic, the mild beneficent acid of milk and cheese and, indeed, of the human stomach.

All fresh wines, which do not have excessive acetic or tartaric acid, contain a bacterium, *B. gracile*, which converts malic to lactic acid, largely in the bottle. This is of great importance to the marginal vintners of Germany, Austria and Switzerland. In the process a small amount of carbon dioxide is also produced, giving a very pleasant, fresh taste. A fresh wine, not quite *pétillant*, is a guarantee not only that the cream of tartar has been removed, but also that no sulphur has been used in the manufacture of the wine. (*Bacterium gracile* will not work with more than one part per 1,000,000 of sulphur present.)

This almost magical process, by which wine becomes naturally less acid in bottle, was always a key indicator of a properly made wine, but the biochemistry was not understood until the 1900s when the *B. gracile* was isolated. There are also other quantitatively unimportant but qualitatively essential acids, such as *citric* and *succinic*, which add taste and character to wine.

In the sunbelt countries – the South of France, Spain, Italy, California, Australia and Algeria – the problem is not acidity, but the reverse. There is too much sugar, and therefore too much alcohol and too little acid. This may be crudely corrected in cheap wines by 'plastering', the addition of plaster of paris or calcium sulphate. This unlocks the cream of tartar (potassium sulphate) which forms as a sludge at the bottom of the vessel containing the wine, whence it may be removed. Unfortunately, although the vintner now has tartaric acid, he also has sulphur in the wine. This is an age-old problem.

Pliny the Elder, probably the most distinguished man to be killed by the eruption of Vesuvius which buried Pompeii in AD 79, had complained two years previously of the practice of plastering, which must have been necessary in southern Italy, France and Spain to sell a wine with less than two parts per 1,000 of sulphate. Even at this level (0.2 per cent) it can be tasted.

So too can the less unpleasant alternative, which is to use calcium phosphate instead of calcium sulphate, thus producing by ion exchange phosphate instead of sulphate. This practice is illegal in the European Union, but it is common in Algeria and in some other Mediterranean countries. The taste has been described as that 'of a wet terrier in the Western Isles of Scotland', and the characteristic damp-dog bouquet used to be disguised (and still may be) in Greece by turpentine, the sophistication euphemised in the word 'retsina'.

Besides the adulterants used in the making of wine, there are innumerable stories (throughout history) of later sophistication, usually by a merchant or salesman ever anxious to give the customer what he thinks the customer wants. The following substances have been added in the past and in some cases still are: new grapes, raisins (dried grapes); plums and prunes; mulberries; elderberries; cranberries (whortleberries, blueberries, bilberries); cherries; black, red and white currants; turnips; mangolds and beets; radishes and grasses – these last two, unbelievably, for the 'stretching and improvement' of white wine.

Sugar (but only refined, cane sugar) has been used for at least 150 years in all stages of a weak fermentation, or thereafter, and the blood of all kinds of animals has been used to give body and colour to a weak, wispy red.

Chemical solutions before, during and after the vintage should never prove necessary for those who apply sufficient care, effort and skill to the avoidance of these problems. Sophistication by merchants (which is still current) is by no means essential. To get everything right has been the answer for thousands of years, and is the most convenient way to create a wine many times the value of a wine which results from any failure at any stage of the work. The value of a wine can be doubled or trebled if the care taken in the making is increased by only a few per cent. And if the wine's crude value is doubled, the profit may be increased many times over. The creation of wealth every year has been a function of the skills and conscientious attitudes of those making wine, and prevention of trouble is always cheaper than any cure. This is as true today as it was 2,000 years ago.

VI

As already noted, the wine-vine is a Eurasian native, with its centre of gravity in the Eastern Mediterranean. It was taken all over the Greco-Roman world before the Christian era, achieving little success east of Persia or north of the Rhine, or in west Britain. It was not at the time taken to East Atlantic islands like the Canaries, the Azores and Madeira. These were visited but not colonized by Phoenicians and became the first wine colonies of the early Renaissance. Wine production was already important when Columbus passed through the Canaries in September 1492.

From the Canaries, Columbus took cuttings with him on his first Transatlantic voyage, but whether they came from Spain or the Canaries we do not know. Like the crew members left behind, the wine-vines planted during the second voyage, on San Domingo, also perished, as did the sugar canes. In the 1530s Hernando Cortez planted in Mexico several strains of vine drawn from the Estremadura by his father, a small landowner who supplied his conquering son with innumerable cuttings, from apples to yarrows. (Apples did better than yarrows.) Mexico was, and is, inferior wine country and the American vine transfers which first impressed the world were in Peru and Chile. The first good Peruvian vintage was in the late 1550s, and the Chilean in the 1580s. Export from both countries started within twenty years, but jealous Spanish merchants at home persuaded the bureaucracy to make inter-Colonial trade in wine illegal. Wine exports had to go from the Colonies to Spain, which often had a surplus. So wine production in the New World subsided to a level no greater than was necessary to look after the needs of priests and settlers. At a later date, wine production, again for local consumption, began in the Argentine, Uruguay, even in southern Brazil. None of it was good wine, until the situation changed recently.

Although South America can now make wine to meet world trading standards, the finest wine in the American continent originally came from California. Vintners in the rest of the United States and in Latin America would probably have to agree.

California has played a crucial role in the wine production not only of the United States but also of Europe. It is the oldest wine-producing area north of the Rio Grande, and makes nine-tenths of US wines today including most of those table wines from the United States that can be compared with the best of Europe. The Californian trade was also responsible for the three worst pests to afflict the vine – pests against which precautions have to be taken even today.

Just as the Church was responsible all over Europe for preservation of husbandry of the vine between the end of the Roman Empire and the Renaissance, so, when the Spanish Conquistadores spread like some thin tide over the Americas, accompanying priests took with them in their baggage cuttings of the European vine, *Vitis vinifera*, which does not occur naturally outside the Old World. This vine was taken by priests via Mexico to California in 1564, and it is probable that small quantities of sacramental wine were made in Lower California, around what is now San Diego and Los Angeles, even before 1600.

The Franciscans later introduced what is now called the Mission grape, in 1767, and small quantities of surplus wine were used to refresh travellers and other mundane consumers from the early 1800s, long before just a few Anglo-Saxon Americans arrived in California. The wine trade had to wait for Americans and others from the East Coast to establish commercial vineyards in their present position in the San Francisco Bay area.

There were false starts. In 1824 a Frenchman, Jean-Louis Vignes, planted 300 acres of vines near what is now the railroad station in downtown Los Angeles. This was unsuccessful on account of sea fogs, which together with the smog still blight the area today. In most years the wine was of inferior quality, so Vignes distilled it into brandy. Before transport became cheap and easy, brandy was a more saleable product than ordinary wine, just as corn whisky was more transportable and often more easily sold than corn. Vignes traded brandy northwards to Monterey and southwards to Lower California. In 1849, the Gold Rush established San Francisco as the centre of gravity in the new State, which joined the United States the following year. The miners and traders, harlots and labourers demanded wine, and within a few years vineyards had been established in the Sierras, Sonoma and Napa Valley. The wine was not very good. It did not have to be. In a Gold Rush people will drink anything.

The key figure in the increase in quality was a Hungarian, Agoston Haraczthny, who settled in Mission Valley, north of San Diego; he planted European vines, which he had imported directly, but the grapes failed to ripen because of summer fog. He moved himself and his vines to Crystal Springs, south of San Francisco, where the same grapes failed to ripen for the same reason. He moved to Sonoma in 1857 and established, or re-established, the Buena Vista vineyard. Before his death in 1864, he had established the Buena Vista Co-operative, and tried nearly seventy varieties of European grape. He wrote a standard work, *Grape Culture, Wines and Winemaking*, proving that vines from European grapes could be grown in California without irrigation. But by sending American vine cuttings to Europe he unwittingly caused nearly as much damage as the War between the States.

California had no native vines (*vitis*) of any sort, but east of the Rockies there are four species: *Vitis Labrusca* (which produces grapes with a 'foxy' taste); *V. riparia*, from the river basins; *V. ruprestis*, from hills and mountains; and *V. berlandieri*, from Texas. These wild species had lived undisturbed by man until the arrival of the Europeans.

Each plant has its own natural enemies. Thus each plant has a particular race of mildew and other fungi which attack it, and is host to certain aphids. Each plant supports certain insects. When mankind travels and promotes and profits from plant transfer, he also spreads the various pests which inhabit those plants. Even if he only moves the fruits of the earth, and not the plants which produce the fruits, he may well move the pests with the product. Thus it was with the three major pests of the vine, which almost certainly came from North America and did enormous damage to the European wine industry.

An early import of grapes into Europe from California introduced white mildew, a fungus, in about 1851: it became known as *oideum*. The leaves of affected vines withered and dropped, the grapes were poor in taste and the wine made from them had to be drunk within a few months, however careful the wine-maker. In France, production of all wine fell by a quarter, in Germany by a third, in Italy by half. Production of fine wine was nearly eliminated throughout the world. Madeira ceased to make wine altogether between 1852 and 1860. *Oideum* wiped out the crop, and sugar cane and pineapple were temporarily substituted for vines in the Canaries, as in Madeira. In Europe generally, wine production was reduced by over 200 million litres a year, at a time when every litre was worth at least $1 wholesale in today's dollar value. The pest was not conquered for nearly twenty years and total loss was nearly $4 billion in today's money value.

In the end, the cure was simple. About 20 kg per hectare (16 lb per acre) of Flowers of Sulphur, distributed round the vines, prevents the mildew as surely as a fungicide sprayed on a cereal prevents the worst effects of the particular mildew which attacks grain crops. Before sulphur was identified as the cure, however, many expedients had been followed, and one attempt at a cure for *oideum* introduced a far worse pest into Europe, Africa and Australia.

During the twenty years before sulphur was discovered as the cure for *oideum*, every conceivable expedient had been tried. One was the bright idea of Agoston Haraczthny, sage of Buena Vista. He suggested the import of *Vitis riparia* and *Vitis ruprestis* from the Eastern United States into Europe. Although these American vines are resistant to *oideum*, they are also hosts to the aphid known as *phylloxera*.[9] The vine leaves, when attacked by this aphid, turn yellow and dry, and die. Without leaves, no plant can live or produce fruit, in this case, grapes. If some grapes are produced, the immature pinheads do not swell; at harvest time they remain hard, small and dry. *Phylloxera* is not like *oideum*; it does not reduce the quality and quantity of wine. *Phylloxera* devastates. There is, simply, no wine. *Oideum* caused a loss of 200 million litres a year for twenty years, but *phylloxera* was responsible for a loss of five times the quantity for twice as long.

California was wholly planted with European rootstocks when *phylloxera* appeared from the east, perhaps brought in a ship or on a covered wagon. However it travelled, the scourge of the aphid was as serious in the 1860s and 1870s in California as in any country in Europe. Taken unwittingly to Europe in about 1860, *phylloxera* did immense economic damage there. In France, production halved. By 1890, the pest had reached every country in Europe, North and South America, Africa and Australia. Wine, in surplus in the 1870s, started to rise in price as it became more and more difficult and expensive to produce. The cure for *phylloxera* took nearly forty years to discover and perfect.

Before an aphid can be controlled, its life-cycle must be known. Aphids of roses or cereals are migrants, moving in waves in dry weather, on the wind. They are destroyed, in nature, by heavy rain or by ladybirds or other predators. Nowadays, they can be controlled by insecticides for up to a month at a time. In the latter half of the nineteenth century there were no effective insecticides available, and in the case of the aphid *phylloxera*, the only predator, analogous to the ladybird, is a mild mite called *Tyroglyphus phylloxera*. It lives in the United States in the native habitat of the *phylloxera*, and its own population rises and falls with the food supply available, as with all control predators. Unfortunately, the theory of biological control failed when the mite was imported into France because it found other sources of food, and the aphid and the mite were to be found living in perfect harmony alongside each other. This is proof – if proof were needed – that man's use of biological controls, so beloved of theoretical ecologists, often breaks down in practice because the control parasite finds other food supplies, in new environments, when moved by man. The deliberate movement of insects is no more guaranteed to be successful than is the deliberate transfer of plants, and for the same reason – that no one except the inspired can predict unintended side-effects.

In the end, sulphur was found wanting; drowning the vineyard proved ineffective; an intensely poisonous chemical, carbon disulphide, was found too expensive. Ultimately, the cure came from the same source as the disease – America. The bright idea was to use the American species, which are *phylloxera*-resistant but which produce indifferent wine, as rootstock, and to graft the superior European species, *Vitis sylvestris*, in various forms, as the cultivar. Today, almost every vine of any consequence in the world is a graft of *V. sylvestris* on to an American rootstock, which confers a characteristic immunity to *phylloxera* on the rest of the vine. This particular technique of grafting took years to develop successfully, and has led in turn to apples, pears, oranges, apricots and peaches specifically grafted to resist certain diseases. Much modern grafting technique which, in effect, can be a form of plant protection or a form of cloning or genetic modification, has derived from the wine industry's response to the scourge of *phylloxera*. Nevertheless, in the forty years it took to perfect grafting, losses were terrible and can be quantified at about $80 billion in contemporary values, nearly ten times that figure in today's dollar value. There are a few, very few, pre-*phylloxera* wines left in the world, and a tiny area of ungrafted vines. Extreme wine snobs affect to be able to claim that the wine produced from ungrafted stock is more interesting than the majority of wine produced on American scions.[10]

California was also responsible for the third great pest of the nineteenth century, the mildew *Peronspera viticola*. This causes leaf-drop; the grapes never mature, but remain sour and soft with a high acid content, since without efficient leaves the plant cannot turn carbon dioxide into sugar. This mildew was, in fact, controlled within five years of its appearance by a discovery made by Millardet of Bordeaux. Spraying with copper sulphate ('Bordeaux Mixture') was instituted as a method of control as early as 1880, and the blue of the vineyard became as much part of France in the decades before 1914 as the blue of the poilu's uniform coat.

*

California was wiped out as a wine-producer by *phylloxera*, and had barely recovered when Prohibition struck in 1919. Prohibition did many things for America, encouraging gangsters, disrespect for the law, and an increase in the consumption of alcohol. But it also produced a boom in legal wine-making. By a deliberate loophole in the Volstead Act, every householder was allowed to make 200 gallons of wine a year at home. Of course, some made much more and, when Prohibition came to an end in 1933, wine-making at home grew rather than diminished, probably because of straitened domestic finances during the Depression. Vast areas in California were planted with grapes to produce grape-juice for the home vintner to make into wine, and for other consumers to drink unfermented. A favourite choice was Thompson's Seedling, a grape with four markets. It could be sold fresh for the table, or dried into raisins, or pressed into grape-juice, or be made at a winery into a strong (over 14 per cent alcohol) clean, unsophisticated and unsubtle drink for winos. This was the position until the end of the Second World War. Only a few, a very few, Californian wines were mentioned fifty years ago in the same breath as the best products of Europe.[11]

VII

As one writer put it: 'The 2000 vintage in most of Bordeaux was of a quality able to look that of 1949 in the eye, and it is unlikely that the current decade will see another vintage to equal either.' No doubt, but the 1949 vintage was handled in a manner which medieval vintners would recognize, while the 2000 vintage was often made with almost revolutionary methods. The 'revolution' in method over the intervening half-century occurred because of the need to meet the requirements of the market, crude financial realities and, above all, the corrosive influence of inflation. Neither war, nor revolution, nor unstable government has mattered as much to the wine-maker as inflation, whose malign arithmetic has altered method and brought chemistry to the aid of the industry in a way which the legion of (mostly nineteenth-century) chemists could never have imagined.

Nearly everyone is aware of mechanical field developments since 1950: tractors in place of horses or oxen; implements instead of men; trucks replacing carts; road tankers substituted for railway wagons loaded with wooden barrels. But the net effect of these mechanical improvements in productivity (perhaps a factor of three) is dwarfed by the contribution of the biochemist (a factor of two in the field; four after harvest). It has been estimated that a vintage of the late 1990s, grown, picked, crushed, fermented, husbanded and bottled by the methods of 1949, would cost many times more, in real terms, than it already does, and that would be to the end of the first eight months only. Thereafter, because of the higher price of storage and credit, to keep a harvest of the 1990s for drinking seven to twelve years later would cost – who knows? Four times the 3 per cent a year compound of the 1930s? The proposition became unthinkable. Inflation, whatever can be done to defeat it in the field, ruled in the cellar in the 1960s, 1970s, 1980s and for much of the 1990s. There was no longer much commercial sense in aiming at a vintage only drinkable in fifteen to twenty years' time. If inflation has really been defeated, then a return to age-old methods may take place; but, crudely put, the need to beat inflation has meant that tannin was diminished, together with much desirable acidity. Unfortunately these characteristics make for an undrinkable young wine but are the very qualities which make the same wine memorable twenty to thirty years later and give it fruitiness, flavour and the incredible subtlety of a great red which no other drink on earth can match. The great problem of the late twentieth century was how to combine the virtues of an old, mature wine with the need to produce the same in half the traditional time. The manipulators are involved at every stage, from the field treatment of vines, the picking of grapes and the fermentation of must to the disciplines of the cellar in order to meet the basic requirements. These are that costs must be controlled by chemistry in such a way that a fine wine can be produced at a price which is not only credible but also leaves a fair margin.

The prospect of standard (but not fine or great) wines being made anywhere in the world, out of grape-juice grown anywhere in the world, and sold to anyone in the world who can afford the cost, is no longer a dream but a projection of what is already current. For this revolution wine-makers are largely indebted to California. This large State, inhabited by more people than any other in the United States, produces more ingenuity, more gross income, more Nobel Prize-winners and more clever inventions than most sovereign nations. California has been called a state which can conceive of at least one new religious sect a month for most of a century, and it is not going to be defeated by the more mundane problem of how to vanquish a cost–price squeeze in its vineyards.

In order to stay competitive with less expensive producers, and to meet – some would say anticipate or even, perhaps, form – modern taste, some wineries have developed the no-hassle method of vinification. This, briefly, is to ferment at a very low temperature, using yeasts (containing enzymes) bred to operate in a high concentration of inert gases, so that of the sugar in the grape-juice more can be turned into alcohol, and less into carbon dioxide. Alternatively, it is now possible to ferment immature grapes into bland, drinkable wine, without risk or trouble. Either way, there are no nasties in the must, and taste can be added later if necessary. All the grapes can be mechanically harvested on any convenient day, regardless of the maturity of the grape or indeed of its sugar-content. Thus grapes can be machine-gathered without problems, and the whole of any area of a vineyard can be cleared, like a cornfield. Fieldwork is much simplified, since the same area, which would normally have to be picked over by hand during a period of several weeks, can now be machine-gathered at a tenth of the cost of hand-harvesting. And it is essential to remember that the no-hassle fermentation technique is vital to the process. Without bio-chemical control of the complex chain enzyme–sugar–alcohol, mechanization in the field leads only to inferior wines.

Many Californian wines are not inferior to the best in Europe. To compete, Europeans have had to charge more for the same quality, or to emphasize the virtues which the Californians cannot claim. Europeans engaged in the enterprise of selling the idea of Old World wines in competition with New World advances call it 'marketing'. They have to use every trick in the book. Territory (*terroir*), grape-type and 'strict' rules of vinification are called into play. At best they are probably likely to be defeated by manipulation in the laboratories. At worst, and if public taste follows the pattern of the 1980s and 1990s, Europeans are likely to have to imitate the New World wine-makers. Some European wine-producers have already flattered the New World by importing its techniques. Others have invested in the non-European wine industry, both for direct profit and as a way to acquire know-how. Other New World wine-producers are already employing Californian field methods, just as the rest of the world adopted combine-harvesters for grain crops, which began to be used in California more than a century ago.[12] Privately, no European quality wine-producer would dare denigrate the success of the New World, whatever they may say in public. New World wines are a success.

Nor can there be another example of a natural product of world quality being achieved within two generations in California and Chile, and in one generation in Australia and in a dozen years in New Zealand. Today, New World wines are a tribute to the know-how, ingenuity and faith of a few hundred people determined to achieve what most Europeans would have said was impossible in 1950. What was probably equally unforeseen by the most optimistic New World vintners in that year was that their world would become, by 2000, the source and reservoir of modern methods, both in the field and in the winery.

From a marketing as well as a manufacturing point of view, Californians have made much of single-variety wines. The law is such in California that since 1983, a varietal wine must be made from 75 per cent of the declared grape. Cabernet Sauvignon, a red-wine variety high in tannin and body, has been spread all over the world by vegetative means (cuttings or layering, not seed). The variety is believed to have altered virtually not at all since Roman times, was very likely in use in the Burgundy and Bordeaux areas

2,000 years ago, and is probably the most widespread of all red-wine grapes. Today, it is naturalized from Chile to Bulgaria; from Australia, through California, to Azerbaijan on the western shore of the Caspian. Although the Cabernet Sauvignon vines have to reach a certain age for the resulting wine to have subtlety (the roots must get deep into the subsoil in order to produce the micro-flavours), and sun is an essential for good yields, the grapes are small and the yield less than is the case with other red-wine varieties. Yet a wine made from 75 per cent of Cabernet Sauvignon vines is rarely a bad wine, as indeed people all over the world have successfully proved – Californians above all.

Varietal wines nearly all made from one grape have been marketed with success by other Americans, from other States, for at least a generation: Gamay for a 'young' Burgundy-type; Pinot for 'Champagne', Chardonnay for white 'Burgundy', Riesling for hock-type whites. Partly, the Americans started using grape-types as generic or varietal titles to avoid derived, inferior names such as 'Claret-type' or 'Burgundy-style'. The ploy succeeded. Thousands of American and other wine-drinkers now think in terms of the variety of grape rather than the territory of origin. The Italians, who are blessed with a comparatively new DOC law which gives their wines territorial integrity with nearly all the advantage of the French AC law, find that single-variety vines may be as good, or even better, as a marketing ploy. There is one prime example from Italy: the Italian red with the greatest future is believed to be Sassiciaia. This was judged in its fifth year of production the best wine made from Cabernet Sauvignon (80 per cent in the case of Sassiciaia) out of a field of thirty-five wines from eleven countries.

Cabernet Sauvignon and its cousin, Cabernet Franc, also have a marked ecological advantage in Italy, California, Australia and other countries. Rich, alluvial, flat (and therefore cheaply worked) land planted with Cabernet vines can, counter-intuitively, produce quality, not quantity. The easiest way to give the lie to the Burgundian saying, 'If your land weren't so poor, it would not be so rich', is to plant Cabernet, which turns rich land into a bank-able fine-wine-producing proposition. Many, though not all, other varieties on the same flat, easily mechanized land would produce dull wines. In marketing terms, more and more vintners will in future probably identify their wines of quality by grape-type rather than use the laws which allow wine to be sold as Bordeaux or Chianti or Rhine-style wine.

It is very easy to call the European love of territorial origin – *terroir* – a mere marketing ploy, but half a century ago Europe produced nearly all the world's great wines and probably 90 per cent of all the fine wines. European wines dominated the trade and, in particular, the fine wine trade. New World wines were often cheap, and rarely distinctive. For those who seek to contest this, the answer must be to examine an international wine merchant's catalogue of 1950 and compare it with one of 2000.

California was the first New World wine province successfully to combine clever, economical modern husbandry systems with the necessary technologies to raise quality without increasing costs in proportion. Today, other New World wineries have learnt from Californian patterns, above all of thinking, and have adapted them to their own conditions. These adaptations now edify every New World wine country, whether in North or South America, South Africa or Australasia, if not every winery.

One indicator is that since 2000, in all wines retailing at more than £5 per bottle in the United Kingdom, the New World was comfortably ahead of Europe. Such a position would have been ridiculed in 1950, if anyone had been unwise enough to forecast such an outcome. But there are those who point out that at the top of the market, at over £15 per bottle retail, the reverse is true. So it follows that Europeans should make the depth and subtlety of Old World wines more widely available in the medium-price range. This is where the blandness of many New World wines should not be at such an advantage; especially important for the new wine-drinker, not yet a connoisseur, who is often apparently intrigued by the artless vinosity of many New World wines.

Great European wines have usually sold at a comfortable profit because of their complexity and subtlety. This has sometimes been due to the blending of two or more varieties of grapes, but also because the same grape may produce several different wines if the vines were grown on (say) three different kinds of subsoil.[13]

VIII

The argument 'Nature versus Nurture' is very much in evidence in all forms of viticulture, but in wine-making the controversy becomes the contest between *terroir* and technique. The Old World uses *terroir* to explain the unique quality of its wine; the New World specifies improved techniques in the new countries' wineries when defining the excellence of the product. Many Old World vintners, especially the French, are professionally snobbish; the New World wine-makers, especially the Australians, are often scornful of Old World methods. But the wise of every Continent judge each situation on its merits and award credit where credit is due.

In some ways, the argument about *terroir* versus technique is contradictory, because every vineyard, regardless of location, necessarily has its own *terroir*. But what is done with the vineyard is what matters – modification must necessarily also involve the other half of the contradiction. *Terroir* can obviously be materially altered by the vintner's actions regarding soil, mineral status, variety of grape, training of vines, water supply and so forth. The advantage that New World wines have over those of Europe is that New World vintners can choose land within wider limits than are available to Europeans. The soil can be tested for mineral deficiencies which can then be corrected or plotted at intervals using GPS technology – accurate to within a metre or two. If irrigation is needed – with or without fertilizers – it can be installed in a green field site much more easily than in an ancient European vineyard dating from medieval times, or earlier. The best irrigation technique, and the least wasteful of water, is a drip-feed system which can be combined with local probes to take account of differences in soil-type and water requirement. Drip-feed is also the most elegant form of irrigation to use for adding chemicals, if and when needed.

In many of the New World vineyards, excessive, even over-active growth can be a real problem which can be exacerbated by irrigation. Such excessive vigour is sometimes prevented in the Old World by expensive hand-pruning through the season. In the New World, the position and spacing of vines are chosen to prevent excessive growth, which can be controlled not only by pruning but also by interplanting crops like mustard or clover – the latter if nitrogen is in short supply. A valuable by-product of interplanting is that insect-eating birds are attracted to the aisles between the vine-rows and eat pests which would, if unhindered, spread disease or eat vine leaves, thereby diminishing plant efficiency.

Without irrigation, it is likely that up to half of New World vineyards would be uneconomic, but irrigation is now banned by the European Union in vineyards. It is true that over-irrigating generally can be as dangerous as drought, leading to formidable explosions of pests which multiply in increased humidity. But some New World vineyards would not prosper without careful irrigation. This is why drip-feed irrigation is favoured, which of course combines well with probes that indicate water-requirement with great accuracy on a continuous basis. The optimum for most crops and most soils and certainly for vines is to maintain a soil-moisture deficit amounting to the equivalent of a heavy storm or about twelve hours' steady rain, say about ¾–1 inch. This policy keeps moulds and fungi, lovers of high humidity, at a safe, low level, and allows for the odd thunder-storm without over-watering the soil. Ironically, a really well-drained soil is even more important when irrigation is employed, since no plant enjoys a waterlogged top-soil; all vines benefit from ground where water drains away rather than lying for a time.

*

The New World has been the cradle of automatic irrigation, mechanical planting, mechanized training, mechanical pruning, mechanized trimming – the last sometimes twice a month in season. Then, before the harvest, there is mechanized leaf-picking prior to the final mechanical task – mechanized harvesting. None of these chores is uniquely mechanized in the New World or wholly disavowed in the Old, but mechanized harvesting is the most controversial.

Some European regions rule out mechanized harvesting and insist that every bunch of grapes be cut off the vines by hand. At the same time, it can be argued that if a vineyard is planned from the start for every task to be mechanized, there is less need for hand-work, even to improve quality. Even the *vendange*, beloved of poets, students and wine-buffs, can be more efficient and far

less laborious if done by machine. But for an efficient mechanized
harvest, the work begins much earlier. For example, probes that
control soil moisture at the optimum level make it much easier to
ensure that every grape contains nearly the same optimum of sugar
and water. If by chance a bad lot of grapes reaches the crusher, its
origin can be pinpointed because the position of each and every
load of grapes can be identified by the GPS system, which can plot
and position every vine within a few feet.

The mundane design of a modern New World vineyard will
have beneficially followed the precepts of the best sort of work
planning: that every product should be made in the most elegant
and economical way possible. In the manufacturing industry, this
is called production engineering, but the same precepts, even
though untitled, are common enough in modern agriculture and
horticulture. But such an ambition is much easier to achieve in a
Neo-Europe than in the Old Continent.

This planned intellectual approach is very different from
the haphazard planting of vines, usually in place of another,
unprofitable crop. This casual policy was adopted many times in
many countries before about 1960, when some of the modern
viticultural weapons became available, trickle irrigation being
among the first. Without these weapons, it is unlikely that New
World wines could have achieved the important position they enjoy
today, where they are beginning to more than rival the best of
Europe.

*

The other end of the wine equation is the vast buying power of
the supermarkets in First World countries. The vital role of grape-
type is partly a tribute to the qualified ignorance of the modern
wine-buyer, no longer a connoisseur, more of an enthusiast keen
not to buy plonk at an exaggerated price. The supermarket array
corresponds with the restaurant wine-list, enticing less than well-
informed but sensible buyers to select the second or third cheapest,
so as to avoid the consumption of utter rubbish, but not to spend
more than is necessary.

Supermarkets have refined the art of offering 'bargains' every
week – usually a variety of different bargains; and this conjures up

a vision of many huge containers waiting to be unloaded and distributed to hundreds of stores on the same day, to be fair to the punter. But so well regarded are the credentials of a wine-merchant, store or supermarket that recommendations are accepted most of the time. It has to be said that Cabernet, Chardonnay or Sauvignon are much easier names to remember than the fancy titles of thousands of vineyards. The New World vineyards are undergoing a greater degree of change and development than the much older vineyards of Europe, the uncontested leader in fine wine production in the earlier part of the twentieth century, with France the world leader in red wine over a span of nearly 200 years. Grape recognition is an important, relatively new factor that informs both wine-maker and consumer, who can be 13,000 miles and several years apart. Even the intellectual French are adopting grape-names if *terroir* does not ring the marketing bell, as in some of the huge vineyards of the Hérault.

*

In the recent past, the transient wine-maker has become a feature of the New World – calling, in Canning's vivid phrase, on the New World to redress the balance of the Old.[14] But transient viticulturists are not concerned, as Canning was, with politics, more with method. They fall into at least two categories. The first is the 'flying' consultant who, like other consultants, breezes in, looks at problems and breezes out again, operating in a cloud of omniscience, part of a carefully crafted stock-in-trade which includes a hefty bill for services rendered. The vineyard owner can spend a long time figuring out whether the exchange was worthwhile. People without imagination often claim that consultants, including wine-making experts, frequently seem to peddle what later appears to be obvious common sense. The recipient is often left wondering why the obvious did not occur without the exorbitant cost of the consultant's expertise.

One, but only one, example of the stock-in-trade of the viticultural consultant is the technology of canopy management. This is a relatively new discovery whereby yields can be increased by as much as 25–40 per cent without any loss of quality. Enough leaves are obviously needed to create sugar through photosynthesis, but too much shade reduces both number and quality of grapes by blocking out the sun. This was less important previously in the Old World when pruning, leaf removal and harvesting used to be done by hand, than in the mechanized vineyard. But excessive leaf cover produces fewer good bunches, uneven ripening and much lower yields.

An open pattern of leaves allows air to circulate properly, which in turn tends to prevent the growth of moulds and fungi, both of which rely on damp, still air to multiply. A careful arrangement of leaf canopy also favours next year's harvest as well as this year's, since buds for the following year are formed by the August of the current year, as with apples and pears.

It is relatively easy and does not take long for a consultant to arrive, show the vineyard owner how to control his canopy of leaves to optimum advantage and move on in an aura of goodwill. The consultant's recommendations, if followed, would in theory make the vineyard far more profitable on the basis of bottles per hectare, and allow owners to recover quickly from the shock induced by the consultant's bill.

The other form of flying wine-maker works harder and longer than a mere consultant. He or she may spend the southern summer in South America, South Africa, Australia or New Zealand, and the northern summer in Europe. Several otherwise undistinguished Eastern European wines, from Hungary and Bulgaria in particular, have been transformed by New World practitioners. Here the New World has transformed the old by the use of example and technique. In particular, some countries have benefited from purpose-built or rebuilt wineries, with control of temperature by refrigeration, some Eastern European vineyards being in areas with a climate as torrid as any in the New World. Given a cooler environment, the next stage is to use a tailored (or bred) enzyme to operate in the best possible way to take account of local conditions. There are other innumerable tricks of the trade, some of them making modern wine-making, for all its technological image, appear to be as much of an art as a science. In fact, if individuality is the objective in a New World (or an ex-Communist) winery where standardization and economic production were once the preferred objectives, the wise wine-maker must achieve specific, even personal, excellence in the product. At the same time it is arguable that until good, standard wine can be made, it is undesirable to try to upgrade the product with elements only derived from personalized talent. This adage may be equivalent to the (flawed?) maxim that until rules can be learnt and instinctively followed, they should not be broken.

Two other points. Until recently, no New World alternative to wooden barrels for the maturing of wine had been found; barrels made of oak, new or once-used, were favoured. Chardonnay, in particular, has a great affinity with oak, but cheaper than an oak *barrique* (225 litres and $750) is the use of genuine oak chips, which are said to produce the same result at a fraction of the cost. This policy also allows wineries to use and re-use stainless-steel barrels with new wood chips added, quantity and quality tailored to need.

The other modern problem is the cork. In many ways, plastic corks are a better bet than the seals of natural origin provided by Portuguese cork trees since bottles were developed in the Renaissance. The use of natural corks generally involves 1–2 per cent failure, producing the disgusting taste of corked wine, as well as economic waste. It is probable that most bottles will eventually be fastened with a plastic cork, even (in cheaper wines) with a metal closure.

The sad side-effect is the probable end of the cork forests of Portugal, which are home to a wonderfully diverse ecology, their loss analogous in side-effect to the arrival of planned, planted forests instead of natural ancient woodland. In place of trees of great diversity of age as well as species, a planted forest is more efficient, but the earth beneath is usually nearly barren of both fauna and flora.

Fine wines (and sometimes great wines) are now made in Australasia, North and South America and South Africa. No vineyard in any country outside Europe produced great wines for the market fifty years ago and probably only in California could fine wines be found outside Europe in 1950. Now there are plenty of fine wines made in the Neo-Europes, and great wines can be found by those who seek them out. They are often called boutique wines.

The most recent (and in many ways, the most surprising)

addition to the New World's wine-makers is New Zealand. As a direct result of the United Kingdom entering the European Common Market in 1973, New Zealand's Imperial Preference in butter, cheese and lamb came to a tapered end and agricultural diversification became essential. Fine wine production in New Zealand was unknown as late as 1990; by 2000, there was a gross harvest of 100,000 tons of grapes, producing wine – 10 million bottles in all – of which half can be called 'fine' and some are entitled to the appellation 'great'. No other single country except Chile has such a range of climate from north to south; New Zealand enjoys more sunshine than do most European vineyards, and many of the best New Zealand vineyards are as near the sea as is the Bordelais. It is a wine country with a great future, its wines subtle and often splendid.

But it is also a nation where the sale of wine – any wine – in supermarkets was illegal until 1990; and the sale and use of alcohol have been inhibited for years by social sensibilities. Native New Zealanders used to affect a simplistic, almost puritanical and teetotal culture, which disapproved of the enjoyment of wine and other forms of alcohol. But as early as the mid-1990s, fine New Zealand wines were on sale in the better stores not only in New Zealand but also in Paris, London, Rome and Berlin. These wines were only on offer on merit.[15]

EPILOGUE

The new millennium is exciting (or horrifying) in its possibilities. Thanks to genetic engineering, we are within sight of breeding enzymes which will produce wines of a particular taste, from grape-juice imported from anywhere in the world, made and marketed near the consumer.

Spanish grape-juice could be shipped to Hamburg and turned into 'Burgundy'; Italian juice could be shipped to Gothenburg and turned into 'Hock' or 'Champagne'; Crimean juice could be shipped to London and turned into 'Bordeaux' – the permutations are endless: grapes grown in the cheapest, sunniest places; wine made near the consumer, and to their taste; product tailored to requirements; a triumph for the theories of Adam Smith and the practices of the marketing men.

Thanks to the skills of the genetic engineer, the yeast breeder and the biochemist, the trade is presented with the answer to the problem of too many of the wrong sort of (natural) bottles chasing too few drinkers. Much better to give the drinkers a rational answer to many problems: give them what they never knew they wanted.

This scenario may seem extreme, but here is some evidence.

Americans started spraying vines with synthetic *Botrytis cineraria* (to produce noble rot at exactly the right time) in the early 1960s.

Nearly all food and perfume flavours can now be synthesized by the petro-chemical industry. Almost any smell can be bought, in powder or liquid form, for a few pennies, from the taste of anchovy to the scent of Zara. Thousands of tons of flavourings are used annually by the food industry. How soon will this be the case with wine?

World-famous perfumes can be imitated in apparently smell-alike forms for far less than the cost of the real product. So effective are the imitations that, once on the skin, only a spectrographic analysis or a great 'nose' can detect the fake. What about cheap wine? Enzymes are now being built in the laboratory to block/ unblock certain precursors to disease in plants, including vines.

Genetic engineers have successfully tailored high-sugar, low-fibre carrots, celery and peppers, each about the size of a walnut, designed to be sold as 'snacks' in competition with candy. Grapes could be given more or less sugar, and more or less tannin. Sugar itself can now be tailored to be low in calories, and non-addictive sugar (and perhaps alcohol) are within the horizon.

Grass and cereal genes have been 'rearranged' to reduce their propensity to freeze. Thus barley and ryegrass can now be ripened in the Arctic Circle. The genes in the species *Vitis amurensis*, which resist cold, can be added to a great wine variety to extend its northerly range by several hundred kilometres, or some means of grafting might achieve the same end. It is fairly safe to predict that some or all of these developments will be in use, if not wide use, within a couple of decades.

Notes

1. See Book I of the *Iliad.*

2. It is important to remember that there were no hot drinks, like tea, coffee and chocolate, made with (fairly well) sterilized water before the seventeenth century in Europe.

3. See William McNeill, *The Pursuit of Power,* London, 1985.

4. It was the shortage of tin ores (compared with the much commoner copper) that made bronze so expensive and fixed some Stone Age people in their technology until they were conquered by metal-users. Some, luckier, went straight from the Stone Age to the use of iron, but most Stone Age people succumbed to Bronze Age conquerors, or themselves developed bronze.

5. How often did this happen outside fiction?

6. Many have remarked on the unlikely nature of the Renaissance if most of the population, including most rulers, were illiterate. But literate churchmen were important patrons of painting and sculpture, while the new merchant class was most certainly numerate, if not very literate. The European development of printing (1450–1500) made literacy an absolute essential for all aspirants, whether in Church, State, Industry or Commerce. This is a neglected aspect of the Renaissance.

7. What has happened in Zimbabwe recently is an indication of what might have happened in Western Europe. Ignorant, violent men seize land they cannot till. Without resident slaves, the victors starve. In Zimbabwe, famine follows seizure of organized farms; but for the monasteries, the same might have happened everywhere in Western Europe after the Barbarians arrived from the East.

8. Today 'designer yeasts', bred to operate at a given temperature and used in place of *Saccharomycea apiciulatus* are increasingly employed.

9. Greek: *Phylon* – leaf: *xeros* – dry.

10. One of the few wines made from pre-*phylloxera* vines comes from grapes grown in the few thousand square metres of Romanée Conti, in Burgundy. This superb wine commands a correspondingly high price.

11. There was a new crisis in 1970, when a new form of *phylloxera* – type ∃ – destroyed rootstocks with a lowered resistance. By the year 2000, most Californian vineyards that had been attacked had been replanted – at great cost. There is much breeding (and some genetic modification) to achieve disease-resistance.

12. Early Californian combine-harvesters were drawn by sixteen to thirty-two mules, oxen or horses, and date from the 1890s, the earliest in the United States and in the world, followed soon in Australia. The modern mind boggles at the skill needed to manage such a number of draught animals. A four-in-hand is considered a tough enough proposition, and all the following carriage does is to run along a road, unlike a combine-harvester which has to be matched (in speed or cutting-width) with the performance of the threshing-machine part of the combination.

13. Nowadays, 'blending' may be done after the wine is made, but purists think that blending grapes, even from the same vineyard, produces a subtler end-result than tailoring the product in the cellar. An extreme case is that a commercial *vin rosé* could be merely a blend of white and red.

14. George Canning (1770–1827), as Foreign Secretary in the 1820s, recognized the new Spanish-American countries, successors to the Spanish Empire. He stylishly 'called in the New World to redress the balance of the Old'. Less dramatically, New World vintners have transformed Old World wines since the 1970s.

15. What is historically astonishing is the speed of success, a mere dozen years (at most) for New Zealand wines to get to the top.

RUBBER

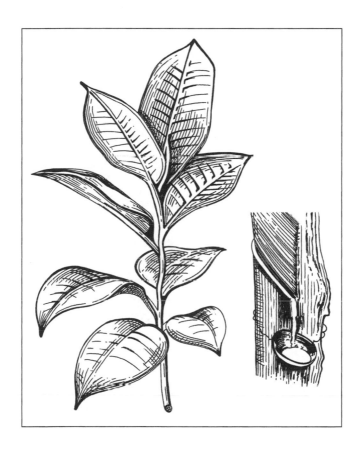

Wheels Shod
for Speed

I

In 1492, most of the world was largely unknown to Europeans: specifically, both Polar regions, most of Africa south of the Sahara, much of Asia, all the Americas as well as Polynesia and Australasia. These 'unknown' territories were arrogantly described as 'undiscovered', but it is fair to say that there were few authentic written records to be found before the first Europeans arrived. So the inhabitants would only know about each other by word of mouth, or song, or verse remembered and handed down from generation to generation. The resources of these areas became vital to Europeans in the next 500 years, a half-millennium of Western European and Neo-European supremacy. Mineral discoveries were exploited at once, but more vital, in the long run, were plants; not surprising, really, since plants grow, while exploited minerals diminish.[1]

Of the plants, rubber was an unknown asset for nearly three centuries, its future importance unimagined. Growing in Iberian Amazonia, the latex-bearing *Hevea* tree needed the French naturalist, Condamine, to notice it and its native use and to paint for the Europeans a picture of what Amazon natives called *caoutchouc*. This name was adopted by the French but the English polymath, Joseph Priestley, styled it India rubber in 1770. This was logical enough at the time because latex was collected by Amazonian Indians and the main English use in 1770 was for erasers. So this enormous, vital industry is now known by the English word rubber for the most trivial of reasons, erasers now accounting for a tiny proportion of the world-wide of rubber, and the better erasers are now of plastic, anyway.

But perhaps rubber was not an absolutely rational name at the time because already, in 1770 in Amazonia, latex was used to waterproof cloaks and hoods, canvas tents and other objects to help people survive in the great rain forest.[2] It was also used instead of candles, there being no fat-bearing animals for tallow in the rain forest (equatorial animals do not suffer from cold and therefore carry little fat). In Condamine's day, rubber was exported via Bélem do Para in the form of syringes or *seringueira*, upmarket water-pistols. So important were these in commerce that the *Hevea* tree became, in Portuguese, *pao de xiringa* (syringe-wood) and latex-collectors *seringueuiros*.

All this only became generally known in Europe after Condamine's voyage down the Amazon in the 1740s, two and a half centuries after Columbus. But some Amerindians from Mexico – Aztecs – had used a crude rubber ball to play a game at the Spanish Court as early as 1531. They had been brought back to Spain by friars anxious to prove their intelligence and skill and enthralled the Court, not only with their skilful play but also with the elasticity of rubber, there being nothing then in Europe that bounced like a rubber ball.

The power of stored energy, one of the qualities of rubber, was of course inherent in bows and arrows. These often-powerful weapons, essential in war before the arrival of gunpowder, were probably known to the Stone Age and, being used every day in hunting, were improved until they were generally abandoned in favour of firearms. But the medieval world knew nothing of rubber, and there was then no other substance like it, no balls that bounced, before rubber proper reached Europe.[3]

The latex-producing tree, *Hevea Brasiliensis*, was to be found wild at a low density of 3–4 every hectare, growing of course at random and being 'tapped' by local Indians who knew which trees yielded well and which were of greater or lesser value on that account. But until much later in history, when commercial export demand grew exponentially, there was little or no competition for the latex, which flowed freely from diagonal incisions made in the bark, the tappers encouraging it to flow into a cup or other container. Today, raw latex is treated with acid to make the rubber coagulate and it can then be easily separated from the residual liquid to be formed into thin sheets. Natural rubber, moulded by hand from latex without any acid, is soft and rots ('recycles') very quickly when exposed to the oxygen in air. Yet years before the discovery of 'vulcanization' by Hancock and Goodyear in the 1840s, rubber had already helped men to fly.

At Versailles, in 1783, a sheep, a cock and a duck became the first recorded aerial passengers, in a Montgolfier hot-air balloon made of silk, coated with *caoutchouc*. The animals flew in the presence of King Louis XVI. Two months later, an intrepid aristocrat ballooned across Paris at a height of 100 metres. Two years later again, the Channel was successfully crossed in a hydrogen balloon. This was two years before the French Revolution, eleven years after the Declaration of Independence, and in Britain the Industrial Revolution had already begun.[4]

Military use of captive balloons for observation followed, but until rubber was stabilized in the 1840s, rubberized silk lost its virtue in the cold high air and practical generals on all sides ruled out their (near-suicidal) use in battle.[5] Important in the American Civil War, captive balloons became essential in the First World War to spot the fall of artillery shells, but they also became a prime target for enemy aircraft.[6] On a fine day, there might be fifty observation balloons aloft on the Western Front, twice as many planes trying to shoot them down and at least four times as many trying to protect them. No balloon or other early aircraft could have flown without rubber.

But the story runs ahead of itself. Napoleon may not have approved of captive balloons, but he believed in fashion; the Empire Line used elasticized cloth, making *décolletage* more attractive than when whale bones alone were used to enhance a woman's shape. Unfortunately, elasticized cloth, widely exploited in both Paris and Vienna between 1805 and 1815, must have begun to stink after a few days as natural rubber, unprotected by any additive, starts to rot very quickly. How many times could such clothes be worn? Or could the smell be disguised with heavy perfume? Other garments were elasticized by the rich: garters, of course, for men as well as women, waistbands, gloves, stockings and headgear. The Prince Regent used elasticized corsets – smell or no smell – from about 1818, but boots named after the hero of Waterloo, Wellington, were made of leather until the 1860s.[7]

Amerindians had made galoshes by pouring liquid latex over their feet, but it needed New England to create an industry of overshoes. Imports of rubber into New England multiplied twenty times between 1826 and 1835 – from 8 tons to 160 tons. Cotton 'shoes' were sent to Para in Brazil to be coated; other rubber overshoes – of excessive size – were imported and lined with cotton or silk. But there was an impermanence about the whole effort, since such galoshes were stiff and inelastic in the cold. Brought indoors, they became sticky and so adhesive that they sometimes proved difficult to remove from hot feet. The very hot summer of 1836 left several New England warehouses a mass of stinking, deformed rubber, fortunately easily turned into fuel. The following year saw an American slump as deep but not as long as the one that followed 1929. This killed the New England rubber industry, with losses of $2 million, a huge sum in those days. There were also said to be 50 million unsaleable pairs of overshoes, many of which were burnt as fuel in the new steam railroad engines. Capital had been invested in six rubber factories in Massachusetts, three more in Connecticut, ten outside New England, one as far south as Philadelphia. There was even a factory on Staten Island in New York Harbor.

Charles Goodyear (1800–1860) is the American hero of the next phase. After seven or eight unsuccessful efforts – he was five times bankrupt – he developed a mixture of sulphur, white lead and raw latex, combined in a so-called 'masticator' invented by a Londoner, Thomas Hancock. The masticator heated the rubber and mixed it with the chemicals, producing a stable product by means of 'vulcanization'.[8]

Contemporaneously with Goodyear's experiments in America, Charles Macintosh in Scotland discovered that naphtha, derived from coal-tar, could stabilize latex into a liquid and make possible a kind of varnish which was applied between two layers of cloth to make a waterproof garment. Macintosh prospered, once he had stabilized rubber, giving his name to millions of garments which were usually made of rubberized silk or cotton and sometimes were lined with wool for warmth. Thus another aspect of the rubber industry came into being, but in America Goodyear died in debt, having earned and spent about $2 million – $50 million today – mostly on patents and lawyers. This was in 1860.[9]

In the previous decade, one great beneficiary of Goodyear's work had been an Englishman, Stephen Moulton, now less famous than Goodyear or Macintosh. Moulton bought two disused woollen mills in Bradford-on-Avon, Wiltshire, and built up a huge industrial rubber plant, the first to provide railways with rubber buffers to reduce shock; Moulton also made other rubber artefacts: springs, washers and gaskets for steam engines, and hoses and flexible tubing.

Stephen Moulton had been on a non-commercial visit to the United States in the spring of 1842. There he met Goodyear, who entrusted him with three samples of Goodyear-cured rubber. Back in London, Moulton met one William Brockendon, a young man of independent means who was obsessed with finding a way to make an effective rubber bottle stopper. Brockendon gave Moulton three addresses at which to leave the Goodyear samples. One of these addresses was that of the great garment manufacturer, Charles Macintosh. Macintosh in turn handed over the sample left by Brockendon to his partner in Charles Macintosh & Co, Thomas Hancock. Spending a year or more on experiments involving sulphur and heat of various levels, Hancock finally cracked the problem by mixing rubber with molten sulphur at a temperature above 240°F. Combining rubber with sulphur made a plastic mix which was not sticky, was impervious to temperature changes and could be formed into any required shape. Applying heat to the manufactured article – sheet, shoes, tube, or whatever – 'vulcanized' it, making it impervious to any further change and resistant to oxidization, of immense use worldwide.[10]

The rubber industry boomed in many places after vulcanization had made it possible to stabilize the product. Between 1851 and 1881, before electricity or the pneumatic tyre, rubber from latex 'hunted' by Amerindians was exported largely from Para near the delta of the Amazon. Total world supply, including non-Para rubber, increased from 2,500 tons to nearly 20,000 tons. (Non-*Hevea* rubber came from ninety different cultivars, but the product

was never really as good as that from *Hevea* exported from Para.)
Variations in supply and demand led wholesale prices for Para
rubber to fluctuate between 38 US cents and $1.25 per pound
during these thirty years. There were, of course, other sources of
rubber as well as many other vegetable products derived from the
sap of trees.[11]

II

Over the next three decades, 1880–1910, rubber became the most
important, most market-sensitive, most sought-after new commodity in the world. Safety bicycles with rubber tyres were selling in
their millions before 1895. The 'safety-cycle' – 'safe' compared
with the penny-farthing – sprang almost fully-formed from the
imagination of engineers: steel tubes, ball-bearings, variable speed
gears, hub and rim brakes, high-quality chains, wheels formed of
spokes in tension – all these were available by the late 1890s.
Bicycles were originally fitted with solid rubber tyres, which always
had the advantage that they did not puncture, a common hazard
when nails were routinely shed from horseshoes. The disadvantage
of solid rubber tyres, however, was not only an obvious lack of
suspension but also difficulty in steering, which became even more
marked in early cars which ran on solid tyres. Solid rubber tyres
for bicycles were always a far more important market than solid
rubber tyres for horse-carriages, though the absence of noise with
rubber-tyred carriages was considered an advantage.

The bicycle was a very favourable investment – cars did not
appear in comparable numbers until several decades later. An adult
could use the same machine for forty to fifty years, the only expense
being a few tyres and other spare parts. With the network of
railways, as it then was, an Englishman with a bicycle enjoyed a
hitherto unknown degree of liberty and mobility, since almost
every train would carry his bicycle as well as himself. The bicycle
was so much cheaper, so much more efficient than a horse that
had to be fed, groomed, saddled or harnessed. But it was only the
pneumatic rubber tyre that made mass cycling popular, and cycle-
tyres were mass-produced ten years after the safety-cycle by Dunlop

in Birmingham, England, Michelin in Clermont-Ferrand, France, and Pirelli in Milan, Italy. Significantly, there was no great cycle-boom in the United States, probably because roads outside metropolitan areas were inadequate and distances too great, and in American cities, public transport was too cheap and potential cyclists too prosperous.[12]

Nor was the United States in 1895 a great place for cars, but fifteen years later, in 1910, it produced 200,000 cars, more than the whole of the rest of the world put together. In 1920, there were 12 million cars registered in the United States and about 2 million built in that year, and more than half were Model T Fords. In the United Kingdom or France at the same date, there were nearly 60 per cent more motor-bikes than cars, a testimonial to poverty induced by the First World War. In even poorer post-war Germany, rubber was in such short supply due to foreign exchange shortages that many goods vehicles, and even some cars, ran on solid wooden wheels, with steel tyres, as in the last days of the war.

The explosion of numbers of motor vehicles, in favourable circumstances, is illustrated by the bus and taxi position in London. In 1900, there were two 'experimental' omnibuses, one steam, one driven by internal combustion. There were five 'experimental' cabs. The other 7,000 cabs and 3,000 buses were all horse-drawn. In 1914, there was not a horse-drawn bus left in London and only a few cabs, mostly used by tourists, as on Central Park South, New York, today. Greens of the twenty-first century should note that the internal combustion motor was welcomed because it reduced pollution, there being nearly 40,000 horses at work in London, each emitting 20 litres of solid effluent a day, or more than a quarter of a million tons a year in all. At least 25 per cent of this, or 70,000 tons, was dropped in the street and had to be picked up, largely by hand.

There was also a great quantity of smelly, high-ammonia urine, which may have reached the Thames by gravity but, on the way, offended not only the sensitive. Then there were tens of thousands of cubic metres of methane emitted by 40,000 horses every day, significantly more damaging to the ozone layer than car exhaust. The use of motors also reduced the number and severity of street accidents since, though cars were primitive, the best, like the Rolls-Royce, were infinitely easier to control than carriages drawn by animals with minds of their own. Even cheaper cars were safer than a spirited pair of horses in average hands.

Finally, modern Greens should note the great reduction in cruelty to animals; omnibus horses, for example, went to the knacker after an average of only two years. Diminished cruelty to horses did not of course apply to Allied armies until after the First World War, and in the case of German and Axis armies, not until their demise in 1945. But *in extremis* on the Western Front, the British used London buses and the French Paris taxis in transport emergencies as early as 1914, to span the gap between the railhead and the battlefield. Both British buses and French taxis had internal combustion motors.

Before returning to the increased supply of rubber, without which no one could possibly have sold millions of bicycles and cars, there was another industry that would depend on rubber – electricity.

In 1860, when few buildings – but including the Gare du Nord in Paris – were lit by arc-lights, coal was the primary source of energy in the First World except, curiously, in the United States, where timber was still king. There was relatively less water-power used than either before or after 1860, steam having widely replaced water-power before that date and hydro-electricity only becoming an important element in the early 1900s. By 1920, there had been an enormous rise in the amount of energy available for a more than doubled world population – fourteen times as much, or six and a half times per head, much of this being coal or water-power turned into electricity. Most energy use in 1920 was, of course, in the First World, the majority of other countries still being dependent – as medieval Europe had been – on the direct use of plant-products, animal power or animal fats, or on wind or water.

By 1880, commercial electrical lighting was fairly widely used – notably for arc-lights ashore in large open spaces, lighthouses and new ships, but there was virtually no domestic use. Incandescent electric lamps had not been developed; coal-gas – in use for street-lighting since 1811 – was more convenient indoors than candles or oil lamps. But gas mantles were only invented in 1885, when domestic electricity had already made a tentative start. These mantles were followed five years later by mantles for oil lamps, which were good sources of light for those living outside the geographic or economic range of gas or electricity. Those without gas or electricity amounted to 80 per cent of the UK population – more in the United States in the 1890s.

All electrical insulation at this date was derived from *Hevea* rubber or other plants that yield latex. Gutta-percha was cheaper for low-voltage telegraph wires, and by 1880 there were submarine telegraph cables, sealed with pitch, that reached all over the world from London, the most famous being across the Channel from Dover to Dunkirk and the Transatlantic cable from Cornwall to New Brunswick. But insulation was often avoided, if possible, so that Edison's first electrical system, in Pearl Street in downtown Manhattan, used bare wires held rigidly taut in the street but insulated indoors.

Early electrical efficiencies were, literally, terrifying. Pearl Street, the first district lighting system in the world, had an overall efficiency of about 2–3 per cent, about one-tenth of what can be achieved today. Almost no electric light installation before 1914 was more efficient than gas, measured as *lumens* produced per ton of coal, but of course there were great safety advantages. There were other advances besides lighting; electrical traction made poss- ible deep Tube trains, early London, Paris and New York under- ground systems having been steam-hauled and shallow. In factories machines could be individually driven instead of by long, dangerous belts, while on the domestic front vacuum cleaners followed early in the new century, as did other appliances driven by fractional horse-power motors between or after the two World Wars.

All these developments needed great quantities of insulating material, meaning, from the 1880s, rubber. In that decade, with negligible manufacture of bicycle or car tyres, exports of Para rubber rose from less than 13,000 tons to more than 28,000 tons, thus more than doubling in a decade. It would be fair to reckon that electricity took 10,000 tons of the increase and, while elec- tricity was a major user in the next decade (the 1890s), it was the 7 million bicycles that existed world-wide in 1895 that needed far more rubber.[13] In the first decade of the next century, the number of motor vehicles with pneumatic tyres rose from a few thousand (at most) in 1900 to a million world-wide in 1910. Demand for rubber took off exponentially, since in 1910 tyres lasted only a few thousand miles and were an expense greater than any other in owning or running a car, apart from the driver's wages (if there was a paid driver). Unbelievable as it might seem to a modern car driver, inefficient tyres, not especially non-skid and subject to blow-outs and punctures, might last only 2–3,000 miles and were never recycled at that date.

Between 1880 and 1910 the three great developments wholly dependent on rubber – electricity, bicycles and cars – increased the demand for rubber to a near-doubling every five, then every three years.

British municipal interest in gas undertakings either took the positive form of active participation, or followed a negative defensive strategy designed to protect municipal gas undertakings from the march of electricity. In the United Kingdom, this led to a peculiar situation that caused there to be about sixty new electrical undertakings in Greater London in 1920, compared with six in carefully planned Prussian Berlin and eight (in competition) in the more Darwinian New York City.[14]

In London, there were several different voltages, and even different AC phasing, in the early days, and there was great confusion about AC/DC which, before 1920s Hollywood, had no sexual connotation.[14] People who moved house usually had to leave behind their electrical equipment, since it was unlikely that the new house would be electrically compatible with the old. It was not only different districts in London that had their own discrete systems; the Savoy Theatre was the first to be electrically lit – by 824 incandescent lamps in 1881 – but by its own generator. The Savoy Hotel next door had another generator (and a different system). So did many other hotels and theatres, for whom the change from gas to electricity removed an important fire hazard. The AC/DC argument was about utility, distance and insulation: DC was cheaper, and the current could be 'stored', up to a point, in batteries; AC had many more power-applications, did not lose nearly as much in transmission, and could be exploited in higher voltages (but required better, more protective insulation). The young Anglo-Italian Sebastian de Ferranti built an AC power station at Deptford in 1891 that supplied power anywhere within twelve miles – virtually most of London then – without significant line losses.[15] But because of fissiparous distribution, his sales were limited.

Another Anglo-Italian, Pirelli, invented an advanced sort of cable, which also minimized line loss, before he made tyres. These cables were available towards the end of the 'electrical decade' of the 1880s. Pirelli ultimately made some very good bicycle and car tyres, but he had cable factories up and running in four countries outside Italy by 1912. Yet another Anglicized Italian, Marconi, developed – as everyone knows – wireless telegraphy, the technique that was the father of speech radio, the grandfather of TV and the great-grandad of cellular phone technology, in all of which Marconi's company played its part, and still does.

It is attractive to draw a pattern from the lives and achievements of these three Anglo-Italians, each of whose work was dependent upon rubber, though no one should claim that without rubber none of them would have left his mark upon the world. Italy had little modern industry before bicycles, electricity and cars. The political union, the Risorgimento, was only completed in 1871, and economically Italy was a collection of city-states, trade and political rivals for 1,000 years and each town much more of a focus of loyalty than the new nation. Italy had not been united since Roman times, and then only under the heel of the City of Rome itself. As an example, after 1870 the Northern Establishment, which then controlled Italy, sacrificed the durum wheat-growers of the south in favour of cheap imported pasta for northern workers – shades of the British Repeal of the Corn Laws forty years previously.

In an industrialized Western Europe, Italy in 1900 seemed notably poor and overpopulated, grateful for the heavy emigration of its people, not only to the United States. Hydro-electricity was Italy's first native source of energy after wood; it was Pirelli's cables and tyres and Fiat's cars that gave Italy more than craft industries and a thriving export trade. It is tempting to identify Italy as one of the countries that rubber made rich, and the development of the new industries made feasible by rubber generated as much positive potential as in any other economy in Europe. It is more than a pity that this opportunity was dissipated by Mussolini's Fascist, ultimately disastrous ambitions. His incompetent, often brutal regime was originally a short-term response to a poor-quality, not very adequate or efficient democratic Government.

Early heroes of the modern rubber industry were not all Italian or Anglo-Italian, however. John Boyd Dunlop was a Scottish vet living in Belfast, and he invented the pneumatic tyre in 1888, largely at the behest of his seven-year-old son, who complained that while riding his tricycle on solid rubber tyres he was more exhausted by the shaking than by the exertion of pedalling. Ironically, some time after the Dunlop Company had been formed and, by then, making thousands of bicycle tyres each year, an earlier patent of 1846 was discovered, as were some other similar patents. André and Edouard Michelin of Clermont-Ferrand called Dunlop tyres *saucissons*, and proceeded to invent the now familiar inner and outer rubber tubes. Dunlop tyres originally had only one balloon that held air, the surface of which ran on the road, like the modern tubeless tyre but with far less efficiency. The *Système Michelin* (of inner and outer rubber tubes) was extended to cars in 1895. So suspect were *pneus* that no competitor in the Paris–Bordeaux–Paris race of that year would fit Michelin tyres. So the Michelin brothers built their own car, not a very good one, but it came seventh out of the nine cars that finished, compared with nearly a hundred that started. What impressed those who travelled in the Michelin car was the smoothness of the ride; what impressed onlookers was the silence apart from the clatter of the motor, a clatter that could only be avoided with an electric car. For the first time in history, there was no sound of iron wheel-rims on the road. It had been two men from West Germany, Gottlieb Daimler and Carl Benz, who separately 'invented' the motor car driven by an internal combustion motor. They both had a few running in 1887, but by 1895 there were *none* in the United States, only a handful in either the United Kingdom or Germany, fewer still in Italy but several hundred in France, the true pioneer country, building in 1895 most of the world's automobiles, whether steam, electric or propelled by an internal combustion motor. There was less enthusiasm elsewhere. In Switzerland, automobiles of all sorts were banned altogether in 1895, while in Britain, until the next year, there was a speed limit of 4 mph, with automobiles preceded by a man holding a red flag.

III

As the 1890s wore on – and the more so in the first few years of the next century – the savagery of the hunt for Amazon rubber and the whole 'modern' hunter-gathering industry struck many contemporaries as being commercially unsustainable. But a man with the solution had been working at a sinecure job in the India Office in 1869; previously he had been responsible for the transfer of the *cinchona* tree from the Andes to India, bringing quinine to the plantation world. Now Clements Markham recommended the transfer of the *caoutchouc* tree from the Amazon to the Far East. In 1870, he had written of the increased demand for India-rubber – this was before electricity, bicycles or cars – and his forecasts about world shortages were wholly based on the need for rubber in steam engines, at sea, ashore, on trains and in factories.

Ten years previously, Markham and the Royal Botanic Gardens at Kew had transferred quinine production from a similar hunter-gatherer regime in the Andes to a plantation economy in India and South-East Asia. Markham now sought to do the same for rubber, and if he had not taken the initiative in 1870, plantation rubber would probably have been unavailable for the car boom after 1900, Para rubber having been only just adequate for the previous cycle craze combined with the great growth of electricity.

Without plantation rubber, it is likely that wild *Hevea* trees in the Amazon region would have been destroyed by excessive exploitation, as were Congo latex-bearing plants by 1913, while Indian *Ficus elastica* also became near-extinct because of over-exploitation. Without the transfer of *Hevea* to the East, there would subsequently have been few cars in the world, very few running on rubber and most – as in Germany after 1918 – on wooden wheels, with a maximum speed slower than that of a fast-trotting horse.

In an extreme might-have-been situation, without rubber eventually available in millions of tons, far, far more than could

ever have been hunted in Brazil or elsewhere, there would have
been little motorization, a few (probably only military) aircraft,
and no electrified appliances in the home. The horse would still
be king of the road, holidays would be taken at home, land-travel
would be by steam train, and most women would still be tied to a
non-electric house and to daily shopping in an unrefrigerated,
non-air-conditioned store, supplied in turn by horse-drawn transport. Historians are rightly contemptuous of the might-have-been
syndrome, but it is a reasonable device to use to point up the
essential nature of a commodity.

Clements Markham, who was later knighted, moved in the
1870s with the deliberate speed of a Victorian civil servant.
Although he was determined to solve the rubber problem in the
same time that it took to establish quinine production in India, it
was actually over a quarter of a century before plantation rubber
made any serious contribution to world supply. The first *Hevea*
seeds reached Kew, then Calcutta, in 1873. They died. A second
batch, in 1875, also expired. Then chance intervened. An English
adventurer, Henry Wickham, a near-bankrupt resident of Brazil
who had tried trading and was about to plant a ranch near
Santarem with *Hevea*, was asked to find 70,000 *Hevea* seeds.
Knowing that collecting seeds was only half the problem, in May
1876 Wickham, used the UK Government's name and credit to
charter an empty, idle British sea-going cargo ship – then berthed
at Santarem waiting for orders. Collecting 70,000 seeds upstream
from Santarem, Wickham moved quickly to avoid the difficulty he
knew existed with *Hevea* seeds: if rained upon, they either germinated at once or would later refuse to germinate at all.

Packed in banana leaves and then in split-cane boxes, the
70,000 seeds reached the steamer without, providentially, having
experienced more than a shower – a meteorological phenomenon
and almost a rarity in the Amazon region.

Then followed the circumstances that some Brazilians have
turned into political legend. Wickham ordered the steamer downstream to Para, where the ship, with a very light cargo only a few
pounds in weight, was seen to be 'in ballast', its screw half out of
the water. Wickham then called on the Provincial Governor.
This was a social evening, but Wickham told the Governor that

the ship 'carried a few important botanical specimens for delivery to Her Majesty's Botanical Gardens at Kew'. The British royal connection appealed to the then Brazilian Imperial Governor; besides, Henry Wickham was possessed of much convincing panache. Nothing was said of the nature of the seeds, or their prodigious quantity, or anything about any intention to establish plantations halfway round the world. The steamer set sail while Wickham was drinking coffee, or whatever, with the Provincial Governor, and he later caught up with the sea-going vessel by tender in the huge estuary of the Amazon river. This so-called subterfuge of Wickham's has led some Brazilians to talk of British Imperialism having robbed them of their birthright – the rubber plant and the wealth it subsequently generated in other countries.

There are three points to be made about this claim.

First, every plantation crop then grown in Brazil came from overseas; sugar came immediately from Europe, ultimately from the Pacific; chocolate, that had once made Para smell so sweet, was brought – by Europeans – from Central America. Europeans also brought coffee from Africa via Europe; cattle, of course, came from Europe, there being none (nor pigs, nor sheep, nor goats) in the Americas. Even the pasture grasses for cattle and wheat for bread came from Europe, as did many food plants beneficial to all the Americas. The only crop indigenous to Brazil and of much trade value was timber, amongst which were the trees that gave the country its name (these trees were virtually exhausted by 1850) and of course the *Hevea* tree. There were also a few spices and nuts, including tropical Brazil nuts. South America would be a much poorer continent without the inward migration of plants, animals and people, as would Europe without American flora.

Second, when the *Hevea* tree was at last grown *en masse* in plantations in Brazil by no less an economic hero than Henry Ford in the 1920s, the crop was devastated by a killer mould to which, at that date, there was no antidote, organic or non-organic.

Nine species of *Hevea* – other than *Brasiliensis* – are identified as being useful for outcrossing, a policy that strengthens desirable genetic qualities, since *Hevea* is inclined to outbreed in nature while inbreeding actively stimulates lower-plant vitality and tends to reduce latex production. Of the nine *Hevea* species with which plant breeders now play symphonies, only two are commercially viable as pure cultivars and neither approaches *Brasiliensis* for yield of latex. None of this was known to Henry Wickham when he collected his 70,000 seeds. But by one of those chances that advances history more convincingly than does any alternative like human wisdom, Wickham had collected all his seeds from *Hevea Brasiliensis* free from a pest later identified as South American Leaf Blight or SALB. So, by great good fortune for all countries except Brazil, all plant material developed outside Brazil came from Wickham's SALB-free trees, growing wild on well-drained ground near the Tapajoz and the Madeira rivers, both tributaries of the lower Amazon. He selected seed from mother-trees not because they were free from SALB, of which neither he nor anyone else knew anything at the time, but because they were high-yielding. SALB-free *Hevea* trees yield far better than do infested trees of course, and we now know that SALB can do nearly as much damage to *Hevea*, which is monocultural, as potato-blight does to potatoes. While potato blight killed thousands of the Irish and changed history, SALB destroyed every effort to grow *Hevea* in monocultural plantations in Tropical America.

Henry Ford, aggrieved by allegedly manipulated London rubber markets in the 1920s, planted large areas of *Hevea* near the River Tapajoz on the same sort of land from which Wickham selected his 70,000 seeds. Like other plantation owners in Brazil, Guyana or Guatemala, Ford lost a great deal of money, and his efforts were abandoned when synthetic rubber was produced in quantity in the United States. Although both breeding and 'budding' for resistance to SALB and other pests have been pursued vigorously for fifty years, Brazil is, because of SALB, the least significant source of natural rubber derived from *Hevea Brasiliensis*, subordinate – in ascending order of output – to tropical Africa, Sri Lanka, the Philippines, China, Southern India, Indonesia and Malaysia, all of which are free of SALB. So much for the legitimacy of any Brazilian claim that Henry Wickham stole their national birthright.

Finally, there was the question of labour. Slavery was abolished in Brazil several times, only to be clandestinely resumed before it was finally extinguished in 1885, yet in the Amazon region, where slaves could defect with ease, slavery was always an inefficient means of organizing labour. So the tapping of *Hevea* trees came to be done by debt-slaves, kept in thrall by a formidably corrupt system that made them walk as far as forty to fifty miles a day in the rain forest. They had to buy all their needs (even including imported food) from the *patrão* to whom the collectors (*seringueiros*) were in debt. But the average *patrão* was also in debt, often owing, like the *seringueiros*, debts amounting to a couple of years' earnings.[16]

The lives of those who 'hunted' latex were short, and few if any died debt-free. Before 1900 hunted rubber also came from Central America, where Aztecs and others had known of latex. There was also inferior latex, often mixed with impurities, from many different tropical countries. Every industrialized European country, in those days of unfree continental trade, sought to guarantee its own supply. Germany encouraged hunting in German East Africa, France did the same in French Guyana and French Equatorial Africa, but Italians like Pirelli depended upon the London market, Anglo-Italian commercial relations being excellent before 1914. French, Belgian and Portuguese colonial traders developed a hunting commerce along the banks of the Congo, seeking latex principally from *Funtumia elastica*. The cruelty, brutality and corruption of rubber collection in the Congo basin provided much of the data reported by Roger Casement (1864–1916), the UK Consul at Boma, the chief town of the Congo, in 1904. His reports aroused the indignation of liberal *bien-pensants*, English and American, many of whom benefited from the use of rubber without having previously thought much about its acquisition. What no one can know, nearly a century later, is whether the situation was as bad in Amazonia as on the banks of the Congo, or worse. In both continents, the tappers were paid not much more than 10 per cent of the export value of the rubber, which was in turn much less than its worth delivered to London. Today, no one can tell whether conditions in the Amazon and the Congo were comparable in corrupt brutality.

By 1913, plantation rubber from the Orient exceeded the tonnage 'hunted' worldwide and in that same year Congo supplies came to an end, the latex-bearing plants having been hunted to extinction or slaughter-tapped, as the inelegant trade phrase put

it. Much slaughter-tapping also, of course, had developed in the
Amazon region when rubber demand was great. And it is difficult
to believe that 'hunting' would ever have been a sustainable policy.
Hevea trees in areas easier to reach, near coast and rivers, had
probably all been slaughter-tapped to death by the last quarter of
the nineteenth century, and it is impossible to imagine enough self-
discipline to prevent death for *Hevea* trees in many other areas in
the Amazon basin if *Hevea* seeds had not been nurtured in the
East.

But there was another reason why the world needed plantation
rubber so badly, and why two and a half prosperous nations would
be founded because of plantation rubber. These two were Singapore
and Malaysia, the half-prosperous Indonesia, suffering today from
more than fifty years of incompetent, venal government.

The reasons for the essential failure of the hunter-gatherer
before the plantation included not only lack of sustainability
because of slaughter-tapping but also volume of supply, a limit
having been reached even without it. But there was a more
immediate, surprising reason: cost. In the boom years before the
outbreak of war in 1914, the true cost of Para rubber delivered to
warehouses in London was four times the cost of Straits rubber
that had a sea voyage twice as long. To Amazonia's inability to
produce the huge tonnages needed by the electrical, bicycle and car
industries was added the fact that hunter-gatherers could never
compete on cost.

And this was so even though hunter-gatherers were virtually
debt-slaves who cost their exploiters less per day than Chinese or
Indian workers on plantations in the East. The labour of most
hunter-gatherers is inefficiently deployed after a certain point in
history, usually defined by population density, but in the case of
Hevea rubber, by the slaughter-tapping of the most easily reached
trees, and by the lower cost of plantation rubber. The effect was
that much Para rubber was unsaleable except in boom times; this
made the whole Amazonian *Hevea* enterprise far more unstable
than it had been fifty years before, when exports had been one-
twentieth of what they were in 1900.

The conclusion is that, in rubber, the hunter-gatherer had to
give way to the farmer-grower, as so often previously with other

crops. It was just too expensive to walk more and more miles, even to 'hunt' what was stationary, to gather the product with great difficulty rather than to grow it densely in a place convenient to harvest. This must obviously have been true many times before in history, with wheat, barley, fruit or nut trees or other plants. It was a unique privilege for modern economists to be able to observe the process in place in the twentieth century, another factor that makes the rubber crop virtually unique.

In fact, economists of 1908–13 did not publicly or at any rate very loudly notice the extraordinary cost difference of four-to-one in favour of plantation rubber. Nor did they draw the obvious conclusions about 'hunted' Para rubber. Perhaps members of the pre-1914 European politico-economic Establishment had sensibilities that prevented them from examining debt slavery and 'hunted' rubber. In the same way, few European economists – or trade importers – acknowledge or investigate the debt-slavery that exists today in the West African cocoa trade.

IV

There were three famous men called Hooker in the 1860s. One was the American Civil War general who provided his troops with professional ladies of easy virtue when soldiers were bored in camp and in need of stimulus – hence one US name for prostitutes: hookers.

The other two Hookers were less colourful and more cerebral: both were distinguished botanists as well as fine administrators. The Hookers, father and son, looked after Kew Gardens for forty-four years. The father, Sir William Hooker, first became Kew's Director in 1841 when he was already fifty-six, an age much closer to retirement than it is now. He and his rival for the appointment had been engaged in a contest for over two years. Each of them could call on a duke for support, in an age when dukes were of great consequence; Hooker had the support of the Duke of Bedford, who had a great garden and a substantial botanical collection at Woburn, and whose impressive accumulation of orchids was given to Kew in 1843, after his death.

John Lindley, Hooker's competitor, had written an important report about Kew's future in 1838, which the Government had accepted. He was a favourite to fill the Director's post and was backed by the Duke of Devonshire, another ducal gardener and a great botanist who had collected (with Joseph Paxton) many interesting and exotic trees and flowers which were planted at his Derbyshire seat, Chatsworth. Both gardening Dukes were Whigs in politics, as was the Prime Minister, Lord Melbourne.

In the end it was Hooker who got the job, largely because he would do it for less money than Lindley. Having settled into the Directorship, he reformed and reinvigorated the gardens at Kew and became an active builder of new and beautiful buildings, reconstructing most of the glass-houses which had fallen into disrepair. More importantly for the story of tropical botany, Hooker reinvigorated the plant-collecting policy initiated years before by Sir Joseph Banks.

It was the generally disinterested purpose of English botanists to 'map' the botany of every 'newly discovered' area of the world. The collectors, usually young botanist-gardeners, were often carried about the world by ships of the Royal Navy on their missions. Sometimes a naturalist was attached to a survey ship, as in the voyage of the *Beagle* round the Southern Seas (1831–6) which started Darwin on the road to his great work.

The discovery and disclosure of botanical knowledge did not have an overtly economic purpose, but botany and business were not exactly disconnected. For example, Robert Fortune, after he gained access to China following the Treaty of Nanking in 1842, not only gained for European pleasure-gardeners cultivars that still enhance our flower-beds but was also able to correct European misconceptions about tea-growing, tea fermentation and tea-curing; many of the errors had been extant for two centuries. Fortune's work in China led directly to the establishment of tea gardens in India and Ceylon, countries previously not known for cultivating tea, but whose tea production exceeded that of China fifty years later. Fortune's work also indirectly became of great economic importance for Indian consumers and Anglo-Indian planters and traders. Before his journeys, few Indians drank tea, except for a small minority of Anglicized Indians who imitated the British, drinking Chinese tea. But this was a taste limited to Calcutta and a few other towns with a considerable European settlement.

Both Hookers, father and son, were significant intellectuals, with conceptual abilities outside the narrow range of the average botanist. Joseph Hooker the younger became Assistant Director at Kew and then followed his father as Director after the latter's death in 1865. Both Hookers were knighted. The elder had previously encouraged Darwin and Wallace in their search for the truth about evolution and natural selection. It was Hooker who stimulated the two men to present their theories in a joint paper read to the Linnean Society in 1858, but it was Charles Darwin, the geologist, rather than Alfred Wallace, the naturalist, who published *On the Origin of Species* (1859) and *The Descent of Man* (1871). Darwin is thereby 'somewhat arbitrarily' credited with changing the world's view of the behaviour of humankind and of the natural world and, ultimately, of the link between the two. What did Wallace lack? Ambition, application – or was he just not ruthless enough? Charles Darwin's mild manner hid a determination unmatched by Wallace and many others.

Because Kew's prestige was enhanced by the efforts of Hooker *père*, after a time he was able to dispense with full-time collectors, the last being one Oldham, who had died in Amoy in November 1864, having collected and tagged 13,000 dried specimens from Japan and Korea. Hooker and Kew thereafter preferred to depend on a network of correspondents, some professional, some amateur, all botanists, most also practical men.

The entry of the first botanical collector into Japan was planned for 1857, the year the British Government presented a steam-yacht to the Japanese Emperor. This was only four years after the famous arrival in Japan of Commodore Peary, US Navy. The British gift of the steam-yacht was accompanied by a plant collector, one Wilford. The planned Japanese plant-collection was not a success. Wilford, who had been engaged to spend three years in Japan, jumped ship (or rather countries) and made an unauthorized voyage, by himself, up the Yangtse-Kiang in Central China. He was then dismissed by letter from Kew. The letter must have taken months to arrive and what happened when or if he received it is not recorded, nor is it any part of the history of rubber to ponder upon the ultimate fate of Wilford, who may never have returned from his authorized expedition. Suffice to suggest that he must have been a brave, enterprising man to sail in those days up the Yangtse, the third longest river in the world – its full length was only properly charted in 1984.

Neither Hooker was the kind of Imperialist exploiter of innocent natives that it is the Marxist and post-Marxist fashion to portray and decry. People at Kew had a genuine belief that skill and knowledge liberated mankind from the drudgery of inexpert toil. But there was no shyness in the High Victorian Age about the civilizing virtues of the European mission in general, and of the British in particular. What was resented by 'natives' of every race was British pressure to forswear slavery or to stop polygamy or to abolish cannibalism or to end human sacrifice or to abandon the practice of *suttee* – the burning, in India, of the living Hindu widow.

Imperial Britain took the lead in abolishing all these customs and habits which were the negation of those fashionable modern virtues, human rights. Such native practices were regarded as the wretched vices of the benighted, as evangelical Victorians would put it. Not only did British Imperialism snuff out perceived violations of Christian standards in those territories coloured red on the map, but pressure, active or implied, on other European Imperial powers led to their virtual end by 1900. British efforts before 1900 had been analogous to US efforts today.

Certainly, without a very active Royal Navy the slave trade would not have come to an end in the nineteenth century. Even if, as cynics claim, the long years of often fruitless, uncomfortable ocean patrols proved to be excellent peacetime training, the one-time slaves of the world owe the Royal Navy a debt. This is still true, even if for more than a century before, the same Navy spent much time and effort protecting the slave colonies of the Caribbean and the sugar-ships on their way back to England, usually with France as the enemy. Slavery today survives in some countries in East Africa – a trade out of reach of any navy.

As with slavery, so with the other perceived wrongs amongst the customs of 'native' peoples. With the single exception of making people in the tropics wear unsuitable clothes in great heat, the British Imperial endeavour was far more beneficial to native peoples than is admitted today. One further point needs to be made. Without imposed Western standards of hygiene and, later, of medicine, many of the inhabitants of the Third World would never have been born, nor their parents, nor their grandparents. No one can view a densely populated tropical city today without quaking at the thought of such a conurbation surviving without any European input.

Smallpox, measles and tuberculosis are far worthier targets if an Imperial demonology is needed. But anti-Imperialists, who are not always rational, are not inclined to tilt their lances at disease, lethal as disease may have been. And it was certainly European diseases that inadvertently did more damage than Imperialism.

Before this unhappy development, the Hookers typically assumed, without intellectual dishonesty or difficulty, that what they did would benefit not only Europeans but the whole world in the long run. For example, when the *Cinchona* tree was moved to India by Clements Markham and Kew in the 1860s, no one in the British Establishment thought to make *quinquina* a British monopoly, and the Dutch, French and Germans were very soon growing as much as they needed. Amsterdam became a centre of the trade and the German chemical industry became a leader in the modification of the original quinine medication and, later, in the production of synthetic quinine.

No one in power in London ever had the idea in the middle or later nineteenth century that any foreigner should be denied the right to buy on the open market in the capital. Given gold, sterling or credit, the City of London was open. That was what Free Trade meant and the combination of Free Trade and the supremacy of the Royal Navy meant that almost any foreigner could travel the world and expect the same protection that was accorded to a Briton. The benefits vouchsafed to a subject of the Queen were also available to anyone else, regardless of nationality. So Sir Joseph Hooker would assume, as had his father, that British efforts to classify botanical discoveries or to move tropical or sub-tropical plants to be grown in a more suitable location would benefit everyone, not only his countrymen. It was with a real sense of altruism that those most closely involved had moved the *Cinchona* tree to Southern India. They ultimately made the product of the bark available to every inhabitant of the malaria-stricken Indian sub-continent for 'a farthing a day', and this was an enormous benefit to the sufferers, even if there were many Britons who saw a great economic gain in a labour force no longer incapacitated by the debilitating effects of malaria. The Indian population doubled twice in the ninety years between the Mutiny and Independence, due not only to the availability of quinine but also to European standards of hygiene.

With the transfer of the *Hevea* plant, no direct altruism could be recruited in support. There was talk in some quarters of 'security of supply'. This often meant that key crops should be grown within the Empire, in Canada rather than in the United States, in British islands in the West Indies, in British tropical colonies in Africa or the East instead of in Latin America. But no one was usually denied the product of the transfer of wheat, maize, chocolate or quinine. All were to be had, for money, on the London market. There was no attempt in the United Kingdom, as in France or Germany, to restrict products from French or German colonies to French or German traders.

Until the end of the 1800s, then, the British made no attempt to follow Continentals down the road to heavy customs duties, or quotas, or tied colonial imports. Such a politico-economic policy was supported by less than half of the Unionist Party (and no one else) and advocated by Joseph Chamberlain in the infant years of the following century. It failed then but found favour in the desperate slump of the early 1930s, as Imperial Preference. Even in 1932, it did not appear to help either the Mother Country or the Dominions or Colonies as much as its proponents had claimed it would.

Kew's part in transferring the *Hevea* tree to the East was an action that would create at least two new nations. But it was undertaken for a number of motives, of which the Imperial cause was minor, if ever enunciated. Significantly, anyone could buy rubber – from Para or anywhere else – on the London market and it was a worldwide benefit that as much rubber as possible should become generally available, which in the 1870s and 1880s meant in London. If there were not enough 'hunted' rubber coming forward, then plantation *Hevea* should be grown in areas whence the product, latex, could also be sent to London. How right the British establishment was became obvious later when French plantations in Indo-China and German plantations in German East Africa (later Tanganyika) were tied to the French and German rubber industries in turn. Before 1914, Free Trade was only really a faith in London, so the only free commodity markets were there, except where Amsterdam filled a role for some products surplus to Dutch requirements. But for the finest bargains and the best service, there was no competition. London was best.

Whether these thoughts passed through the minds of those responsible for the transfer of *Hevea*, no one now knows. But in the late 1870s, the *Hevea* plants sent to Calcutta had failed, since they needed a more constant temperature and humidity (without a dry season). Some of the plants sent to Ceylon, and others sent to Southern India and to Burma, had been propagated with little success. Planters in Ceylon itself were bemused by the profits generated by coffee, far greater than those produced by cotton, sugar or tea (the last from Chinese cultivars). In 1869 a coffee blight appeared, producing a vivid orange blotch, and this blight proved to be as devastating as the potato blight in Ireland twenty-four years before, in 1845. But ruin took longer to develop in Ceylon; only at the end of the 1870s did coffee blight comprehensively defeat the planters. They tried rubber, but by chance they chose the wrong *Hevea*, cultivars from Ceara instead of from Para, and the Ceara *Hevea* trees were killed by incessant monsoon rain. For a time, rubber followed coffee into limbo. Temporarily, Ceylon's planters managed with *Cinchona* and tea while rubber had to look elsewhere for its new Far Eastern home.

V

By chance, and only by chance, a few *Hevea* seeds had been sent by Joseph Hooker to Singapore Island in 1877 and planted in that year. In 1888, another great opportunity presented itself: Henry Ridley was appointed head of the Singapore Botanical Gardens. The son of a Norfolk clergyman, like Nelson, Ridley was educated at Haileybury, which had been founded to provide competent professional men for the East India Company. While its role had been much altered after the Indian Mutiny in 1857, the company had been responsible for the foundation of all three British settlements in Malaya: Penang, Malacca and Singapore. Apart from these three towns, all seaports, there was little European development. Malaya in 1888 was devoid of roads and railways, and transport through the thick equatorial jungle was by river. About the only important inland economic activity was tin-mining, often carried out by Chinese entrepreneurs who dredged for the metal, not originally using steam. The Malay Princes, known as Sultans, ruled a patchwork of small states, only three of which at that date were 'Federated', meaning protected by the British and in turn acknowledging the Queen as their suzerain. The others were 'unfederated' and not always at peace. Gradually, the number of federated Malay states increased and the unfederated fell in number, the total remaining the same.

Henry Ridley, who was to do as much as anyone to establish rubber in Singapore and Malaya, had been trained as a geologist, but spent seven years in the botanical department of the British Museum, now the Natural History Museum. When he reached Singapore in 1888, he spent much of his time experimenting with the best way to nurture *Hevea* trees – now nearly a dozen years old – in their Singapore home. He was also much concerned with how best to increase the production and flow of latex without damaging the trees' longevity. In the end, he developed new techniques of a sort that had never, obviously, been needed in the Amazon, where *Hevea* was 'hunted', not husbanded as in the new plantations.

Rubber 'hunted' in the Amazon region, even when a relatively small tonnage was exported, was important to Brazil. In some years, Brazilian coffee produced as much export income as rubber; in most years in the last quarter of the nineteenth century, rubber exceeded coffee in export earnings. The same competition between coffee and rubber was also played out in East Asia in the early days, that is, before 1900.[17]

Henry Ridley's jobs included the cataloguing of the flora on the mainland of Malaya in addition to his primary duty as Director of the Botanical Gardens in Singapore itself. He became very quickly the Number One propagandist for planting rubber all over the

Malayan Peninsula, to benefit the economy in general and planters in particular. He was convinced that rubber, properly managed, could outperform any alternative crop under Malayan conditions and even carried *Hevea* seeds wherever he went, packed in damp charcoal powder in closed containers, which prevented any deterioration. Ridley planted *Hevea* seeds all over Malaya as he catalogued the native flora in the jungle. In his experiments with *Hevea* he was planting for the future, but originally, in the 1890s, planters in Malaya had by and large been believers in coffee, not rubber.

Three men would challenge this negative outlook for Malayan rubber. In November 1895, eighteen years after the *Hevea* seeds had reached Singapore and been planted successfully, the Kindersley brothers asked Ridley for some *Hevea* seeds to plant as shelter from the sun for their newly sown coffee plants. This was on their estate near Kajang, in Selangor, not far from Kuala Lumpur. In fact *Hevea* is not ideal to shade immature coffee plants, which need an umbra less dark than that beneath *Hevea*. Nevertheless this plant introduction in 1896 ultimately led to a successful rubber career for the Kindersleys, one of them, Ronald, becoming an important and weighty figure among the rubber fraternity.

Another early hero of the rubber revolution in Malaya was a Chinaman living near Malacca, one Tan Chay An. His father, Tan Teck Guan, was a distinguished amateur botanist whose correspondence with Kew had led him and his son to meet and become friendly with Ridley. One of the earliest Chinese to emigrate to Malacca, Tan's ancestor had become the official spokesman for the Chinese community in the days when Malacca was still ruled by the Dutch (after 1824 it became a dependency of the British East India Company). The Tan family was also prominent in civic matters and in the promotion of charity. Tan Chay An planted the first serious rubber plantation in Malaya, 48 acres of *Hevea*, the seed supplied free by Ridley from his trees in Singapore. These 48 acres were, in 1896, the first rubber plantation of any size in Asia.

Also in 1896, there was one of those periodic slumps in the London wholesale price of coffee: green Brazilian beans, then considered the best for reliable quality, fell in price from £4 per 60-kilogram bag to 10 shillings (50p). Malayan coffee did not have the quality *cachet* of Brazilian beans, so its price fell even further – it was the sort of marginal product that only sold really well during booms. Then some of the mould-pests that had destroyed the coffee plantations of Ceylon reached the plantations in Malaya. A brisk demand grew for *Hevea* to plant in place of coffee. Ridley, who had stopped giving away seeds after launching Tan Chay An into rubber, started supplying *Hevea* plants, not seeds. The outcome for plants was more certain as well as more rapid.

Now Ridley required the Singapore Botanical Gardens to charge for young rubber plants and the Gardens sold nearly 15,000 of them in the last eight months of 1896. Ridley also supplied (free of charge) good technical recommendations on planting, spacing, husbandry and tapping. In 1897, 32,000 rubber plants were sold; in 1898, four times as many, and by the end of 1901 nearly a million young *Hevea* trees were growing in Malaya. As part of this great new industry, as it became, most coffee plantations were rendered redundant in the owners' views. Acre after acre of diseased coffee crops was ripped out and replanted with *Hevea*, and a thought immediately occurs: if there had been no coffee blight in Malaya, the like of which destroyed the industry in Ceylon, would rubber have been planted on such a large scale? One certainty exists, that Malaya would not have been enriched by growing coffee as much as it was by growing rubber. It is a possible fancy that, with no coffee blight, Malaya and Singapore would be no better off than are, say, the Philippines with its huge population and a far lower average income than that enjoyed today in Malaya and Singapore.

In the spirit of Free Trade, the British in Ceylon had forwarded some *Hevea* seeds to the Dutch in Java; these failed to grow into productive trees, either because they were planted in the wrong ground (too wet) or because they came from the wrong province of Brazil, from Ceara rather than Para, as did so many of the *Hevea* seeds originally sent to Ceylon in 1877. More successful were the seeds shipped from Singapore in 1883, as were some other seeds consigned from Kuala Kangsar the next year. From these two gifts grew the Indonesian rubber industry, now second only to the Malayan but at one time far ahead. In 1900, the Dutch East Indies had more than sixteen times as large an area planted to rubber as was seeded in Malaya – 82,000 acres against 4,700. One reason was topography.

In Java especially, the Dutch had encouraged a plantation economy with primary products, those not consumed in the East or in Holland being re-exported through Amsterdam. This was a legacy of the Dutch East India Company, an organization that had supported the Dutch high civilization of the seventeenth century. The tradition persisted and made the Dutch Empire more commercial, less military or concerned with sovereignty than the British. Some of the East Indian islands were settled by the Dutch two centuries before the British arrived in Malaya. The Dutch grew every tropical crop: quinine, sugar, coffee, tea, palm oil and various fruits and spices – all produced and sold on a market-garden basis – and rubber was added to these. Diversity was an important Dutch colonial objective, since primary products (the 'softs' of commodity markets) were (and are) vulnerable to changes in demand, yield, weather, fashion and currency changes, not to mention disease.

The then Dutch-ruled island of Java had two great advantages when rubber was introduced. Much of the forest land had been cleared, some even before the Dutch arrived, since Java was always relatively more densely populated than other islands in the East Indies; for the same reason, there was always plenty of skilled agricultural labour. So prepared ground, which had formerly been cropped, plus the labour to work it, were both to be found in Java.

Sumatra, like Malaya, was heavily forested and relatively underpopulated. The time and capital investment needed to convert raw jungle to plantation were vast and, if added to the time needed for *Hevea* seedlings to become productive, meant a massive investment, far greater than in nearly any other plantation crop. There were two years of clearing primary jungle, which, in an age before bulldozers and other earth-moving equipment, was back-breaking human toil, not always accelerated by the assistance of elephant and buffalo.

There were roads and paths to be built, and the plantation had to be drained – no small task in an area with a rainfall of nearly 2.5 metres a year, or 5 centimetres a week, so that 500 tonnes of water had to be drained or transpired from each hectare each week.

Then followed two of the most expensive operations: building a perimeter fence to keep out wild animals, which usually included local elephants, and constructing houses for the planter, his family and workers, and buildings associated with the preparation of latex. Much (but not much more than half of this expenditure was saved if a plantation was converted from, say, coffee to rubber, but in total investment in rubber was several times greater than that required with other crops. There was also a much longer delay before any estate became profitable compared with, say, coffee. Together with the need for market-skills alien to most planters, these elements led to the development in Malaya and Singapore of the Far Eastern agency system.

The first two important agencies in Singapore were Harrison & Crosfield, and Guthries. They had very different histories. Harrison & Crosfield, originally in Liverpool and then, from 1854, in London, were basically importers into Europe; they traded in tea but also in spices, sago, coconut oil and almost anything from the Tropics with a market in the West. They had become specialists in tea from Assam, then Ceylon; at times they underwrote capital for tea estates, a role that would be even more important when it came to rubber-planting, with its large initial investment. After planting, the agency would be responsible for the buying of all the necessities of the estate, down to food, and they looked after centralized finance, accountancy, administration and management. Because agencies came to manage several dozen estates, they were able to compare the effectiveness not only of technique but also of managers. The legendary visits of field agents were of necessity usually made without notice, since there were few roads, let alone telephones, before the mid to late 1920s.

After 1885, when a highly intelligent new partner, Arthur Lampard, joined the firm, Harrison & Crosfield adopted a pro-active stance and opened offices in Colombo, Ceylon, Calcutta, Malaya and the Dutch East Indies. Lampard concentrated on rubber after the popularity of the bicycle and later the car had made it obvious that rubber was a commodity that would be far more important than others. It was Lampard who most helped the Dunlop Rubber Company into Malaya in 1901, and in the following year the firm floated its first estate (not Dunlop) on the

London Stock Exchange. Dunlop would become one of the largest estate owners in Malaya and investment in rubber estates would reach more than £500 million before 1920; of these, Harrison & Crosfield were involved in about a fifth.

Guthries, the other important firm, was before rubber a local Singapore agency for banks (usually British), insurance companies and steamship lines and dealt in the export of tropical produce and the import of the necessities of life for the better-off expatriates, whether they were European or Chinese. Guthries only went into rubber in 1906 and ultimately carried out the same sort of agency business as did Harrison & Crosfield and other, lesser firms. The important distinction was that Guthries was Singapore-based, while Harrison & Crosfield remained firmly London-oriented. Guthries encouraged the Singapore market for rubber, which was seen sometimes as a rival to London. The firm also took a much more lively interest in local Singaporean politics and sometimes actively supported the authorities in Singapore against the British Colonial Office, with the result that Singapore was not always regarded as an ally by the authorities in London, but sometimes even as an adversary.

VI

Before car tyres were ever made, the United States was buying a third of all the rubber used in the world every year. Later, even before the First World War, there were times when the United States imported and used considerably more than half the rubber that reached the world market. In such circumstances, it seems strange that the centre of the huge American rubber industry was not in a port but in a small town in Ohio, now a great city.

Akron, Ohio, was built on an Amerindian site of trading importance, albeit not a densely populated area; there were probably several square kilometres for every Indian then alive in what would become the State of Ohio. The site that would be Akron was on the Cuyahogo River, which flows 40 miles north into Lake Erie, and just to the south of what is now Akron are some lakes that drain, ultimately, into the Mississippi Valley system and then to the Gulf of Mexico. Amerindians had used this route, from Lake Erie to the great river, as a way of moving furs and other goods, in canoes or on foot, and they called the place that would become Akron 'Portage' in their own local language. This was long before any Indians acquired horses, which Europeans had brought over after Columbus's voyages and which had gone wild and moved north from Mexico via Texas, reaching the Plains Indians before the first white settlers came west from the North-East. Europeans only arrived in what would be Akron around 1800 and a generation later, there was a small town built on a crossroads, as were so many Mid-West towns.

In 1825, at the time of the pre-railroad canal boom, during which the Erie Canal had been built to join the Lake to the Hudson River, a retired general, one Perkins, bought land at the summit between the River Cuyahogo running north and the Lakes to the south. Not wanting to call the small town by an English translation of the Indian 'Portage' or 'Portage Summit' or 'Summit', which seemed too commonplace, the General trans-lated 'Summit' into Greek – hence Akron. He also determined to build a canal to run north–south and another to run east–west and this was done, to be followed in due course by railroads which ran along the line of the two canals.

Akron, before rubber, prospered modestly, trading like so many other small towns in the Mid-West in grain, corn, hogs and cattle, which were raised locally and shipped to larger cities by canal, later by the railroad. This agricultural idyll was interrupted when the first rubber entrepreneur arrived soon after the Civil War. Benja-min Franklin Goodrich reached Akron in 1871, aged thirty, having answered an advertisement which sought to attract new enterprises to the town. It was George Perkins, son of the General, who had advertised, as the head of the Akron Board of Trade.

With Perkins as his (not always willing) source of capital, Goodrich moved machinery from a failed rubber manufactory at Melrose, NY, to Akron. Then he started to make fire hose, belting and billiard cushions as well as less exotic necessities for steam engines. The B. F. Goodrich Company was incorporated in 1880 but Goodrich, worn out by years of promoting himself and his company, died prematurely, aged forty-seven, in 1888. It was Perkins who raised the Goodrich Company to the heights that it had attained by 1900, when it was recognized as the technical leader of the US rubber industry. It was the Goodrich Company that had earlier made the first US pneumatic tyre for a pioneer automobile manufacturer, Winton, who had read in America about the Paris–Bordeaux car race and Michelin's pneumatic tyres. Goodrich went on to make many fine pneumatic tyres, without reference to either Dunlop or Michelin, leaving the patent situation to look after itself.

The second rubber investor at Akron was the grandiosely named Goodyear Tyre and Rubber Company. This was formed after the panic of 1893 had driven down real-estate values and made available a bankrupt workshop which one Frank Seiberling bought at a knock-down price. In 1900, this volatile entrepreneur hired an engineer-graduate of MIT, Paul Litchfield, who was Seiberling's antithesis, a methodical man, staunch and steady, and a fine factory manager. This alliance of fire and ice would soon turn Goodyear into the largest US rubber manufacturer, with sales in 1920 of $169 million. The following year, the Goodyear Company was nearly broken by the slump and Seiberling was thrown out of the company by creditor banks, but he ultimately started another – smaller – rubber concern in his own name. Paul Litchfield remained in charge of Goodyear, to become President of the company from 1926 until after the Second World War.

The third great Akron rubber pioneer was Harvey Firestone, who, born in 1868, was working in his cousin's horse-buggy factory in Columbus, Ohio, when a stranger arrived from Detroit to look at the solid rubber tyres on the buggies. The stranger was Henry Ford, who had built a horseless carriage; the year was 1895. This first automobile was not a Ford but was built by Ford for his employers, the Detroit Edison Illuminating Company. The car was on bicycle wheels which, naturally enough, were not up to the weight of the horseless carriage, even though it was really a light quadricycle. But Ford had heard that Firestone had a novel and effective way of attaching solid rubber tyres to the iron rims of the wheels of buggies. Firestone on this occasion made some modest proposals about the rims, the spokes and the solid tyres of the wheels for this very early Ford-designed automobile. This was one of the very first horseless carriages made in the United States.

Firestone and Ford later became friends as well as commercial allies; Firestone owned his first tyre factory in 1900, another by 1903, and made his first pneumatics in 1905. Three years later he supplied 'improved' pneumatics for Ford's Model T and became the favoured tyre-maker for this famous automobile, the world's first true mass-produced car. Both men became very rich.

In the 1920–21 slump, which seriously damaged Goodyear, Ford and Firestone met the great loss of value in raw rubber, pneumatic tyres and automobiles by extensively reducing prices to clear stocks. Suffering great losses, albeit temporary, Ford and Firestone slashed prices, but kept their factories employed and their order books full while others had to lay off workers and lost sales in the subsequent upturn.

Ford, Firestone and Thomas Edison used to take summer holidays together, choosing a different wilderness for their camp each year, their vacation including vigorous discussion – almost a seminar – about the future of industry and commerce, the state of economics and technology and so forth. In 1921 they were joined by President Harding and the resulting inevitable publicity about the President and his holiday companions revealed the location of their camping holiday, previously a closely guarded secret. This year, the site of the holiday was in West Virginia and the outing gave the press and public great delight when the clandestine nature of the annual trip was revealed as a foible of the very great. Ford, Edison and Firestone were icons, far more than mere household names, and probably better known and more highly regarded than President Harding, who later turned out to be a deeply flawed man. Besides, in those times the White House impacted little on the public.

In the early 1900s, about the only large American rubber outfit outside Akron, Ohio, was the United States Rubber Company. This was based in New England and was primarily a boot- and shoe-maker, and an important one, selling more goods, in dollar terms in 1905, as footwear for people than all the other manufacturers who shod carriages, bicycles or cars. Following the failure of

the pre-vulcanization overshoes (English 'galoshes') and their foul-
smelling nemesis, it had not been until the 1850s that rubber boots
and shoes again reached the position once held by the manufacture
of unstable rubber footwear in 1836–7. This time, because of
'vulcanization', the overshoes did not stick to the foot or stink to
high heaven in hot weather.

The New England business – The US Rubber Company – was
more adept at hedging the market than the Mid-West tyre-makers
in Akron. While US Rubber only dabbled in tyres in 1906–7,
unlike other rubber manufacturers they rode out the huge build-up
of demand which followed the great rise in the number of horseless
carriages, now universally known as automobiles or cars, and the
slump of the following year.

VII

In 1900, the best ribbed rubber sheet (RSSI) cost CIF New York
about 60 cents a pound. This figure rose, partly pushed by market
manipulation but mainly because of the huge proportionate
increase in car numbers, to 75 cents at the end of 1904, and to
double that figure, $1.50, in the autumn of 1905. No other
markets had known such a quick rise, nor was there much opportunity materially to increase the 'hunted' output from Brazil, whose
ultimate record production was only a few thousand tons greater
than in 1905. Nor was 'hunted' rubber from anywhere else –
Central America, the Congo or India – the answer. Any considerable increase in latex production had to come from plantations in
the East Indies. But, though a great price rise would increase the
planting of new *Hevea*, it would be nearly a decade before any new
acreage, planted in 1905, would significantly affect supplies or
prices.

Two rubber companies had safeguarded their position by
buying forward – the United States Rubber Company, in touch
with markets in Boston and New York, and Firestone, in Akron.
Firestone's fortunate instinct had led him to buy forward, thus
allowing him to meet Ford's need for tyres for the Model N at a
sensible figure, lower than proposed by his competitors. Other

rubber companies in the United States held off the market when their current stocks had been used and there arose something of a drought in new tyres, none of which were made for stock but were only available to order at an enhanced price.

Amidst market manipulation by a group of London rubber merchants known in New York financial circles as the Brazil Pool, at the end of ten months the price of RSSI fell from $1.50 to 70 cents. At this price in New York, many Brazilian concerns found it impossible to compete, but the cost of production in Malaya was equivalent to only about 20 US cents a pound landed in New York or London, so the effect of the near 100 per cent rise and 100 per cent fall in rubber price was to exaggerate the swing from wild, 'hunted' rubber to the cultivated plantation product. New estates in Malaya were floated on the London Stock Exchange when rubber prices were high and, if a low-cost production base could be proved, even when prices fell. Some of the older estates in Malaya and Sumatra had made respectable profits when RSSI was at 30 US cents a pound and a fortune when the price doubled. But only a foolish (or desperate) Stock Exchange operator floated an estate when rubber prices were very low, if aware of what was going on in the automobile business in the United States.[18]

These were exciting times. There was a series of bank failures and a Wall Street slump in 1907, not as long-lasting as that of 1893 and nothing like as severe as those of 1837 or 1929, but bad enough for the Federal Reserve System to be set up to prevent or minimize the graver forms of bank failure.

Recovery, helped by the automobile boom, was rapid. In 1908, the Model T arrived on the streets, Ford workers were to be paid the unheard-of sum of $5 per day, and General Motors was founded. GM in that year made more cars than the whole of Western Europe, and Ford and GM together more than the whole of the rest of the world. Ford's output would increase tenfold in the next ten years, but GM would overtake Ford in volume and profitability before the end of the 1920s.

RSSI having fallen in 1907 to less than 50 cents a pound for a few weeks, rose to over $1.25 in 1908, to $2.15 in 1909 and to over $3.00 in 1910. In that last year, there were two con-current booms going on in London. The short-term boom was in rubber itself – spot rubber, forward rubber and all the rest of it. The other, longer-term boom was in rubber plantations. In 1910 they were floated at several times the numbers of 1905 before the 1906 collapse in the price of RSSI. The Stock Exchange boom in plantations, new and old, was reminiscent of the railway mania before 1845 or the dot.com obsession in 1999–2000. In the United States, the reaction against the London rubber market-makers and London-controlled rubber estates was as lusty as any previous discontent with British Imperialism. Americans took steps to free themselves, once again, from the British. This time the War of Independence would be economic and commercial.

Americans objected to a perceived market-rigging operation in London in a market characterized by some senior British figures as one that New Yorkers could not rig. There were the Malayan taxes, in particular export taxes and a land tax, which might benefit the British colony directly and the metropolitan British indirectly, but could never benefit the American consumer of rubber. Finally, there was a freight cost greater by cargo-liner than by tramp-steamer; this was called 'The Conference System'. An allied consequence was that rubber might be freighted to London or Liverpool, then unloaded and trans-shipped to a Transatlantic vessel to be unloaded again in Boston or New York. There were, in 1910–11, few direct cargo sailings from Singapore to the United States. This last shortcoming disappeared during the First World War when, to avoid German U-boats, rubber was shipped directly from Singapore to San Francisco. Thus raw latex became an important freight for US railroads from the West Coast to Akron and New England.

One response of the Americans to the British stranglehold was to become producers of raw rubber themselves. In 1911, the United States Rubber Company, then still larger than any firm with a head office at Akron, bought 50,000 acres (later increased to 100,000 acres) in Sumatra. Here more than 25,000 acres were planted within eighteen months, more than 60,000 acres before 1917, the whole acreage by 1920. Goodyear joined US Rubber in Sumatra a few years later and these huge areas were worked with local labour, supervised by Americans or Europeans. They became model estates. There were other important US investments in the Dutch East

Indies, notably in oil. Other Americans sought to plant rubber in the Philippines, which the United States had annexed after their war with Spain in 1901, but local land-holding legislation made large estates in the Philippines virtually impossible. Later, as will be related, there were other American attempts to escape from the British Empire.

Other nationalities adopted other solutions to the supply problem. The Michelin brothers planted thousands of acres in French Indo-China, north of Saigon, thus escaping the market for a large proportion of their needs. Other non-British enterprises bought acreage in Malaya: the Danes (in the form of the East Asiatic Company), the Belgians (with French participation) and the Japanese (on the east side of Malaya). None of these investments helped Malaya or its inhabitants very much. Other than the Japanese, who used Japanese labour, these concerns hired Chinese or South Indians to work their estates and all of them tied exports of rubber to their home manufacturers. They became tropical market-gardens for the home factories, but the tied estates only really supplied the basic requirements for the rubber factories in Europe or America and there was always a surplus demand that had to be made up from the market when times were good. Even the largest estate owners among the rubber companies, Dunlop or US Rubber, probably only supplied three-quarters of their normal requirements from their estates. They had to buy rubber in boom times and their requirements would be secured by their own estate production only in years of low market prices for rubber; it is not clear what advantage there was over the manipulation of the market by forward buying, hedging and so forth. Perhaps the owners of these huge estates not only produced rubber but played the market as well. But for every speculator making a gain on the commodity markets, there has to be someone else suffering a loss, even if the loss is only on paper.

The greatest rubber speculation before 1914 occurred not in Singapore, nor in Penang, nor in London, but in Shanghai. This was at the same time as the London boom in rubber shares which reached its peak in 1909–10, but the Shanghai boom was based on shares not really respectable enough for the London Stock Exchange. Some of the 'estates' floated in Shanghai were frankly fraudulent – cleared jungle, not planted, without a perimeter fence to protect the ground from wildlife and with no buildings yet assembled. The boom in Shanghai went on for a couple of months after the collapse of the similar boom in London. Although Shanghai Chinese, like the Chinese in Malaya or Sumatra, remained great investors in rubber, the nature of the boom-and-bust on the Shanghai Stock Exchange in 1909–10 dented any claim that this great vibrant, stimulating city might have to be taken seriously as an important force in the prudent investment in rubber estates. The experience of those two years was too exciting, too dire for the wise to trust what went on in Shanghai. Hopes that the city would play a major part in attracting finance for new rubber shares were killed by the quality of the boom and its apparent dependence on fraud. Shanghai would never thereafter challenge London or Singapore. But the experience of those who made – and lost – fortunes so quickly in those two years was sure, in the argot of the period, to sort out the men from the boys.

VIII

The First World War did not really much affect life in the Far East in a directly military sense. Both Japan and Russia were Allies of France and Britain from the outset and the German colonies, apart from German East Africa, were quickly captured and occupied, and German cruisers swept from the seas. Because Italy was also an Ally after 1915, the Mediterranean was safer than in the Second World War, and the Mediterranean and the Suez Canal remained a favoured route to the East, even though both Austro-Hungary and Turkey provided bases for German U-boats. Altogether, the war of 1914–18 was much more of a European War than a World War. One of the peculiarities of August 1914 was the cry 'Business as Usual'.

But no one could guarantee such an outcome, however well promoted. The rubber auctions and markets came to a halt for a month after the outbreak of war on 4 August 1914. When the markets reopened, prices had not changed much, but then, between October and December 1914, the price of rubber in London and in neutral Amsterdam and New York nearly doubled. Shipments to

Holland increased by enough to suggest that Germany was building a stockpile to defeat the Royal Navy's blockade. The British Government belatedly woke up and banned the shipment of rubber to any country not allied to Britain. This embargo excluded exports to the United States, by far the largest consumer of rubber in the world, buying in normal times more than 50 per cent of all rubber offered at auction. A hastily hatched plan gave the Americans the right to import rubber from London, Singapore, Ceylon or Malaya in normal pre-war quantities. Tyre output in still-neutral America increased enormously as American car production increased between 1914 and 1917, and much of the extra rubber was needed for tyres for extra cars. At the same time, cultivated, Far Eastern rubber supplies rose by much the same proportion, and by 1919 Malaya alone produced more rubber than did the whole world ten years previously.

As already noted, one very interesting wartime development was the direct shipment of rubber from Singapore to the American West Coast. Concurrently, the Singapore auctions, which would include RSSI from the Netherlands East Indies as well as from Malaya, gradually became more important, as price-indicators, than auctions at Mincing Lane in London. This was not really surprising; Americans were buying more and more rubber as their car production multiplied while the Europeans were at war. Some Americans also, it appears, preferred to deal in Singapore rather than in the City of London, where many Americans maintained that they were the victims of the wily British. Even after the United States declared war on Germany and its allies in April 1917, American car production still increased despite wholesale conversion of factories to war production. As in the Second World War, the Americans built a war economy on top of the original peacetime economy and when victory came, they added the war economy effectively to increase the capacity of the pre-war peacetime economy. The size of the US economy in 1918 or 1945 was, say, 150 per cent of what it had been in 1914 or 1939, and should be compared with the shattered state of Europe in each case, with more damage to France in 1918 than in 1945 and much more damage to Germany in 1945 than in 1918.

Some economic damage has always been self-inflicted, and self-inflicted damage usually arises from the efforts of the well-intentioned but wrong-headed. This was the case in rubber at the end of 1918. The US authorities had promoted the idea of a 10 per cent duty on rubber imports as a revenue-raising tax. This was never passed by Congress, but the damage was done. American tyre and rubber companies scaled down their buying because they thought that the port price of rubber would fall. Meanwhile the British, conscious of the general shortage of shipping space because of sinkings by U-boats, restricted rubber imports to about 20 per cent of actual requirements. The US Government then decreed – by Executive Order – that rubber imports should be limited to about half what was actually needed. There was also a rumour that there was a 'secret stockpile' already in America. All this reduced demand at a time when supply in the East Indies was increasing rapidly as new acreage came into profitable production. Thousands and thousands of tons of rubber piled up in store in the Straits and only in the late spring of 1919 was there enough shipping to shift supplies to rubber-starved manufacturers in Europe and America. There was also a sudden rush of common sense which allowed Governments to withdraw restrictions.

The nature of the boom in 1919–20 was to affect not only raw rubber but the Stock Exchange price of rubber manufacturers in the United States, as well as rubber estates in the East. The boom-and-bust was as severe as the Shanghai boom of 1909–10, even if not founded on fraud. The collapse of prices was as bad in America as in the East and, as already related, every tyre-maker was in real trouble, including Henry Ford and General Motors.

In Malaya, the hardest-hit employees were unmarried British men on the rubber estates who were without employment contracts of any kind and were dismissed out of hand. As improvident as the estate managers who had hired them, they often had no resources whatsoever and had to walk to the nearest town. Some of these men had fought as officers in the War and were ill-equipped to deal with a lifestyle close to destitution. No one knows how many were affected, but the true number probably ran into thousands. One cargo line brought home several hundred in one ship, in accommodation normally used by Muslim pilgrims going to Jeddah, the port for Mecca, payment to be made by their relatives back home in England or Scotland.

The human price of the rapid collapse of rubber prices was far worse in Malaya than in the Dutch islands, largely because the Dutch had much more of a plantation culture, while most of the Malayan estates had been hacked out of jungle only a few years previously. Most of the young Britons on the Malayan estates were inexpert, often just out from home, while the Dutch gave the impression that they had been in the islands for hundreds of years and they pursued a more diverse, more traditional pattern of production on their plantations. They had also been neutral in the First World War.[19]

*

In the 1930s, there was a *Punch* cartoon in which one young man-about-town asked another, 'Is this whisky pre-War?' 'No,' was the answer, 'but it soon will be.' In retrospect, the years between the two Great Wars seem to have been made up of ten post-war years and ten pre-war years. One very important difference between 1939 and 1914 was that neither Italy, Russia nor Japan were Allies – a huge difference, not balanced by the neutrality of Turkey or the disappearance of the Austro-Hungarian Empire. Italy waited until France was defeated in June 1940, after which Mussolini behaved like the proverbial jackal and tried to claim the carcass. In the end, the Italian Empire proved to be the carcass, not France or Britain. Nor were there any jackals.

Russia had signed a non-aggression pact with Hitler's Third Reich which gave Stalin an illusion of security in exchange for raw materials that Hitler needed for his war machine. For those for whom Red Socialism and Black Socialism were closer than liberal democracy, there was no great surprise in these circumstances. For those on the Left, for whom Communism had been a great improvement on Tsarist Russia, it was proof that British and French Conservatives were unable to bring themselves to an alliance with Communists. To others on the Left, it was proof that the British and French Governments had tried and failed to set Hitler and

Stalin at each other's throats, a circumstance which many not on the Left preferred, it was claimed, to another Anglo-German war.

In the end, of course, a much-weakened Britain won the Second World War, while Germany, even further weakened, lost it. Russia, perhaps more damaged than either, was yet able, as a dictatorship, to marshall sufficient strength to occupy Eastern Europe and to challenge the British and Americans worldwide for more than forty years after the Peace of 1945. As will be seen, an important part of this challenge was in the East Indies.

Japan is the most curious case of all. The Anglo-Japanese Treaty of 1902 was due for renewal in 1922; for a number of reasons it was not renewed by the British, who professed at that date a romantic view of the utility of the League of Nations and an even more romantic faith that the industrial and military power of the United States would once again rescue the British. The British decision not to renew the treaty with Japan was, in fact, taken after the United States Senate had refused to join the League of Nations, which was further weakened by the absence of Russia. Moreover, the Americans in the 1920s made it quite clear that they had no intention of rescuing Europe again or, as it was put, 'to pull any more European chestnuts out of the fire'. American isolationism was the order of the day and the United States had become, for the first time, a considerable creditor-nation, reversing a relationship that had made London the source of most of the capital that had built up the Colonies, then the States, for 300 years. American bankers had not immediately learned the first rule of conduct for a creditor: allow your debtors the means to pay interest and to repay the principal. This was illustrated by the War Debt question. Britain, which was neither being repaid the wartime loans made to Allies nor receiving reparations from Germany after 1921, was unable to repay the much smaller loans made by the Americans to Britain. American bankers were unwilling to accept anything but gold or US dollars, nor was Congress ready to reduce tariffs to allow British exports to the United States to earn dollars. In the case of some manufacturers, the tariff was designed to be so heavy as to make imports impossible.

The second point of aggravated difference was the question of the British Empire. The United States, founded by thirteen ex-British Colonies, contained many who made a great point about supporting dissident anti-British movements in the British Empire. This was intensified in the administration of President Roosevelt after 1932; it became the unstated war aim of Roosevelt's administration to encourage anti-colonial movements, and some of Washington's war aims in 1939–45 led directly to the dismemberment of the British Empire. In this respect, it is important to remember the influence of the Irish vote in Congress. Irish-Americans led the anti-Imperialist movement, and regarded Ireland as the first (as well as the nearest) victim of British Imperialism. The weight of Irish-American opinion on the Anglo-Irish troubles in 1919–23 was as vital as the same influence on a similar situation since 1968.

All these factors had an important political and economic influence on the rubber industry. Communist Russia conducted its scientific effort and all its foreign trade through Government agencies, but it had been Russian academic scientists who, in Tsarist times, had identified the possibility of making an early synthetic rubber, *butadiene*, from industrial alcohol. Because of general incompetence, this did not achieve success in Russia but did in Germany and the United States. Second, Communist Russians also made fairly frequent claims promising alternative sources of natural latex. There was a dandelion-like plant, *tau-sagyiz*, that had about 15 per cent of a latex-like hydro-carbon in its root. Huge tonnages of this 'rubber' were promised, but in fact little bulk manufacture ever took place. One American, shown a laboratory specimen, thought it looked like used chewing-gum and said that this 'synthetic' was unlikely to reduce Russia's demand for natural rubber. No one knows how many scientists and engineers Stalin's henchmen executed or sent to punishment camps, but some of those working on the *tau-sagyiz* project must have been among the victims. The Communists also adopted two policies in relation to Malaya and Singapore. First, they bought natural rubber secretly (and successfully kept their secret) through local agents, without revealing tonnages. Second, and pregnantly for the future, the Russians recruited Chinese workers into the Communist Party, to act as terrorist 'sleepers' until after the Second World War.

Japan turned away from the West and felt isolated; Britain was no longer her friend but usually now a trade rival, and the United States was a naval and commercial competitor. Japan found difficulty in selling her industrial products in a world that was increasingly protectionist. So Japan ploughed a new Imperialist furrow, turning on China, seizing Manchuria in 1931 and declaring war on China itself in 1937, by which time Japan was clearly in the Axis camp. The latent militarists had defeated the latent peacelovers (and admirers of the West) in Japan's complex culture.

The political situation in Japan after the First World War contrasted with what would happen, under American tutelage, after the Second World War. As far as Malaya was concerned, the Japanese enlarged their successful rubber plantations on the East Coast and made plans involving their military for the future invasion of Malaya, Singapore and the Dutch East Indies. There was an even wider project called the Co-prosperity Sphere which, though known to British military intelligence by 1936, was conveniently ignored by politicians in London, who were neither properly recovered from the world-wide slump of 1929–32 nor resolute on what to do about the threat from Nazi Germany.

Japanese economic plans for East Asia leaned heavily on racist and anti-European sentiment, and made the Japanese a master race, like the Germans. Auspiciously, Berlin and Tokyo were far enough apart to prevent the two from meeting, and their cooperation during the Second World War was not all that it should have been.

One technical point is revealing about Japanese industrial attitudes: the Japanese developed methods of using the latex in lower grades such as RSS2 or RSS3. It was said that Japanese manufacturers could make a decent tyre out of rubber that no European or American would consider buying.

The British Establishment in 1918–19 was much more prepared to contemplate State action than had been the case before 1914. Having experienced effective Government intervention in the manufacture (not only the purchase) of munitions, ships, aircraft, even electricity, the view spread in Whitehall that politicians and civil servants could help directly in other situations. Some wartime bureaucrats, typically those who had come into Government from industry, did not altogether share this faith.

All through 1921–2, the price of rubber fell to between a third and a quarter of a 'fair' price, defined as one that would yield a reasonable return to most Malayan producers. The value of the British investment in Malaya, reckoned to be worth £150–£175 million at the height of the boom, had fallen, according to Stock Exchange valuations, to about £50 million. There was a serious threat that American rubber barons could mount a pre-emptive take-over bid for Malayan estate acreage. There was a well-sourced rumour that a large fund – of perhaps $250 million – had been put together on Wall Street with just this objective; it was certainly not an impossibility, but it turned out to be untrue.

After six months' haggling in London, Singapore, Penang and Kuala Lumpur, and a crucial failure to involve the Dutch, a scheme was set up to restrict output from Malayan and Singalese estates to 60 per cent of their 1919 yield. This restriction was to force up the realized price; but it was opposed by nearly all growers in Ceylon, who had lower overheads than those in Malaya and, more significantly, the Dutch refused to join the scheme. When the inevitable rise in the price of rubber shipped from the East came about, the chief beneficiaries were Dutch growers who were not in the scheme and were able to increase their sales while sales from estates in Malaya were limited by agreed quota restrictions.

There were too many factors working against the success of the scheme to be described here, but smuggling from Malaya to Singapore, sometimes via Sumatra, usually by sea and by night, was a profitable game which netted the owners of small Malayan estates far more profit than they would ordinarily make. The small estates were, of course, able to survive on lower returns because they had much lower overheads than British-managed estates; equally, when prices rose, the Chinese made far higher profits.[20] A scam worth mentioning is that of coupons. The right to export was validated by the existence of coupons, printed by security printers De La Rue. These were forged in Singapore without, it appears, any great difficulty and they were apparently available in quantities to suit any exigency. Finally, there was what was called native rubber – the natives being Malays of the islands, not overseas Chinese. Native rubber was produced in quantity in the Dutch East Indies, not always on a regular basis but usually only if profitable. As in the Amazon fifty years before, *Hevea* trees might be left untapped until prices rose. Acreage followed the same pattern, rising with the price, the area increasing during the 1920s so that by the end of the decade there were said to be a million acres of native rubber. The only cost when trees remained untapped was the expense of weeding.

The effects of changes in rubber supply on any restriction scheme were predictable. But, of all the influences that made the scheme a political failure and, later, a prop for American isolationism, the strangest was the simple error in calculating demand for rubber in the United States. This was authoritatively judged in London to be 300,000 tons a year, while the best internal American industry estimates were for 400,000, rising to 500,000 tons. Predictably, the price of RSSI doubled before 1925 and increased thereafter.

The concern of the Akron tyre-makers and their customers – car builders and owners – was not only about the price of rubber; there was also near-anger that the price of a vital raw material could be manipulated by the British Imperial Establishment. There was a serious proposal in Congress that rubber imports from the British Empire should be banned, while imports from other territories should be encouraged. Harvey Firestone planned, bought and established a large rubber estate in Liberia, which succeeded; he claimed that he could produce rubber more cheaply than anyone in the Straits. Henry Ford planted a large acreage in Amazonia, which failed. In both Germany and the United States, chemical firms took the production of buna, the synthetic rubber reformed from an alcohol base, one stage further than the laboratory. In both countries, potential mass production was postulated to be only a decade away while Dupont reconnoitred and researched throughout the 1920s and in 1931 patented neoprene, dubbed a wonder-material because it most resembled natural rubber. Unfortunately, the first news of neoprene coincided with a year when demand for raw, natural rubber was the lowest in any year between the wars.

In July 1925, US rubber buyers had been down to one month's supply. The restrictions had worked temporarily, not because of their validity but in view of the great rise in demand for tyres and other rubber goods in the United States, unforeseen by British

Imperial thinkers in London. These had consistently underestimated the rise in American need; besides, no one in London
apparently realized that the second side-effect of restrictions on
Malayan estate output would be the end of any planting for the
future. The first effect was, of course, evasion, but the lack of
new planting would tend to maintain the high price of Malayan
rubber, taking into account the time for each *Hevea* tree to come
into profit. As demand grew, this second unintended consequence
would become more and more important.

The British Government was not worried about the new high
dollar cost of rubber, rather the reverse. When New York rubber
was higher than 75 cents a pound, rubber dollar profits accounted
for not only the interest but also the annual repayable tranche of
US War Debt. The Chancellor of the Exchequer in the Conser-
vative Government after the General Election of 1924, Winston
Churchill, had changed the pound sterling to its pre-1914 parity.
This immediately made sterling commodities and manufactures
more difficult to sell, but at that date there was no substitute for
natural rubber, which was the most vital and most valuable export
to the United States.

Pre-war exports had largely been Victorian products. A decade
and a World War later, the chief British exports still numbered
three: coal, dug by human and animal muscle out of the earth and
raised by steam; iron and steel products, made with less modern
means than elsewhere, and cotton spun and woven on machines
that had not changed much in half a century. None of these
products could be profitably exported to the United States, because
of tariffs. In Britain, there were over a million unemployed and
another million men had been killed or maimed in the War, so
there were fewer males in work than in 1914; this factor made
everyone shy of trying to save labour and thus raise productivity.

Even in the new industries, like cars or chemicals, output per
man-hour was far lower than in the United States, in car production lower even than in France, in both cars and chemicals lower
than in Germany. This ratio (output per man-hour or PMH) of
course ignores any difference in hourly wages; what matters more
economically is the wages element in the end-product. The Ford
worker was still making Model Ts in 1924–5, and these were sold

at half the price in 1925 compared with 1910. But Ford workers in the United States were paid twice as much in 1925 even though the wage content of the car was far lower, because the assembly system had become so much more effective in terms of PMH.

This excursion is essential if the stagnation of the British economy in the 1920s is to be understood. There are two other socio-political points to be made. First, more than half the population in the 1920s probably believed that it was workers alone who created wealth. This was the driving conviction of the trade unions and the Labour Party and many others on the Left; these sentiments were shared also by (often sentimental) Tory ex-officers who were conscious of the sacrifices made by workers in uniform they referred to as 'Poor Bloody Infantry'. It was a sentimental view because infantry, improperly armed, trained and led, were obviously more likely to be casualties than winners in battle. Few naked men could defeat an armed opponent nor could primitively armed men beat a well-furnished foe. As with armies, British manufacturing industry was in the same situation, for virtually the first time in history. But there's none so blind as those who do not want to see.

The British Establishment preferred to regard the pre-1914 situation as 'normal', but the world had changed completely. Stupidity had often produced slaughter on the Western Front, through the use of the wrong weapons, the wrong training and the wrong leadership. Now the same sort of institutional stupidity presided over an unnecessarily high level of unemployment.

The second instance of British cerebral failure was to believe that London was the centre of the universe, the British rulers of the Empire, the lords of all they surveyed, and the pound sterling the benchmark for every other currency. In fact, the most important material point about the difference between 1914 and 1924 was that Britain had become a debtor-nation and was poor compared with the United States. This point did not really sink in until rearmament and the losses of the early war years had turned Britain from debtor to near-beggar. By the time of Pearl Harbor, the United Kingdom was broke and its war effort subsidized in every significant way by the United States.

The same institutional stupidity played its part in the rubber story. The British refused to reduce the restrictions because of the temporary benefits of the high rubber price, which buoyed up sterling's new Gold Standard parity and significantly helped to repay the US Debt. But Herbert Hoover, then the US Secretary for Commerce, told a secret session of a Congressional Committee that US consumers had paid $100 million more than they need have done in two years for goods involving rubber, like tyres. Enmity was directed at the British, ignoring the Dutch, who had also benefited.

This story has been told at some length because of its effect on the much larger issue of how Americans viewed Britain, Europe and the British Empire. The British abandonment of the Anglo-Japanese Treaty in favour of a non-existent American alliance was as big a mistake as when, a few years later, the British made ordinary Americans pay twice as much as was fair for tyres in the first country converted to the car. British policy towards rubber greatly swelled Isolationist sentiment, perhaps contributing to the British near-defeat in 1941–2. Rubber restrictions were especially unwise since they offended a class with many interests in Britain and otherwise most unlikely to be Isolationist.

IX

In April 1928, the restrictions on estate production were lifted, not because the British Government especially willed it but because it became impossible to maintain them. The price predictably fell in the next eighteen months, a time of great American expansion with US vehicle numbers about two-thirds of all road vehicles world-wide. The New York rubber price fell from more than $1 to about 20 cents and this was reflected in the production chain. Darwin's survival of the fittest applied and only the most efficient, least extravagant producers rode out the eighteen months of declining price, which were judged tough enough by the survivors, but worse was to come. The break in Wall Street in the autumn of 1929 halved the value of quoted shares but affected commodities even more severely. Not every commodity attained its price nadir at the same time, but at the very bottom, maize (corn) reached 14 per cent of its 1928 average, barley 9 per cent, and rubber fell from 20 cents to 3 cents, 15 per cent of its 1927/8 average, already down from $1.50 only three years before. There were inevitable demands for renewed restrictions on output, if only because estate overheads in Malaya were 13 US cents a pound, but lower in Dutch estates at 10 US cents a pound. Native rubber, by 1932, involved a million acres in Sumatra, and natives could not, it was said, give up production because they no longer grew their own food but had to tap rubber to get cash to buy food. So no restrictions were enacted, nor would they have worked since prices were considered too low to be revived by anything except a true rebirth of demand.

So the agony went on. The slump in demand for rubber was far more prolonged than in any previous depression. (In 1837, the total world rubber trade had amounted to only a few hundred tons.) Now, the effect of the slump on estates was to put many on a care-and-maintenance basis, with labour and management drastically reduced. Tamils were sent home to Southern India, and local Chinese workers dismissed – and how many became Communist in consequence? In the end, after months of negotiation, a new commodity agreement grew out of the prolonged gloom, to take effect in mid-1934. Nearly scuppered once again by unaccountable native rubber, the price rose from 3–5 US cents to 13–14 cents in late 1934, to 15–16 cents in 1935, and to 25–30 cents in 1936. By the last year, the Germans had reoccupied the Rhineland and their anti-Semitism had become vicious, while Japan's intentions became clearer to every wise man.

By 1936, it is likely that Germany and Russia were stockpiling rubber, both also encouraging synthetic. One Scotsman, Sir John Hay, now began to lobby for the United States and Britain to do the same. In the end, the Americans not only stockpiled a year's supply of natural rubber before Pearl Harbor but also built Federal-funded synthetic plants that by 1945 were making over a million tons of synthetic a year. This, and rubber recycling, allowed the Allies to surpass Germany in this industry, vital in modern war. But in the Straits, Europeans, Malays and Chinese were to sustain more than fifteen years of trauma.

The formidable troubles of natural rubber-producers – two slumps and two commodity agreements in twenty years between two wars – should not obscure the great growth in the use of the product. World car numbers doubled between 1919 and 1929, and low pressure 'balloon' tyres – first developed in the United States – needed nearly twice as much rubber as did tyres in 1919. On the other hand, tyres with their treads hardened by the use of carbon black lasted longer. Electrical insulation, hard rubber 'stick-on' soles and heels, soft crêpe soles, waterproof boots, substitution (rubber for ceramic in hot-water bottles being a minor example) and hard rubber (much used before modern plastics) all brought rubber increasingly into the home. To look at refrigerators or vacuum cleaners of the 1920–40 period is to realize how rubber made possible a great advance in domestic appliances. Clothes of all kinds, inner and outer, used more rubber than ever before. Sport was more and more rubber-based, not in only shoes and clothing but in balls, bats, rackets, even golf tees. Many 'Victorian' sports and games derive from a time when vulcanized rubber became available: lawn tennis, football, even billiards. Rubber became essential in hoses and seals of every kind, in suspension of all sorts, in sound-proofing, heat insulation and carpeting. Aircraft were stuffed with rubber, more so even than were cars. Raffles Place in Singapore was relaid with rubber blocks (like the private road outside the Savoy Hotel in London) in the Depression, a use again encouraged in the 1950s when natural rubber appeared to be challenged by synthetic.

Perhaps the greatest benefit to many was the successful use of rubber in birth control. From the 1920s, 'rubber goods' became widely available, though not yet in slot machines. Even then, there was a gender gap. The male sheath could be bought at any large pharmacist in most large towns, though its sale was still illegal in some States in the United States; but the female equivalent, the Dutch cap, needed careful fitting and a consultation with a doctor or nurse. Nevertheless, the availability of fine rubber contraceptives, much promoted by Dr Marie Stopes and others, was of great assistance in birth control, even though it produced another gender gap. A man could put on a newly bought sheath with slick panache and give his partner an illusion of safety, even in the back of a car. For a woman, the fitting of the analogous contraceptive contrivance gave her an uncertain degree of control over her own fertility,

a control that only the Pill would make nearly 100 per cent real in
the 1960s.

X

After Europe went to war again in 1939, East Indian rubber
producers could not predict the future but they were to sustain two
periods of extreme pain, grief and peril. The Japanese assaulted
Malaya on 8 December 1941, as they did Pearl Harbor. They sank
two British battleships two days later and gained control
of the local seas. Penang was occupied on 18 December, Kuala
Lumpur on 11 January and Johore on 31 January. Singapore fell
on 15 February 1942 to a bicycle-mounted army half the size of
the British Commonwealth force. The British were ill-equipped
for the Tropics, ill-trained and ill-led. Churchill said that it was
'the worst disaster and the largest capitulation in British history'.
Not long afterwards, the Dutch East Indies were occupied.

The Japanese had not signed the Geneva Convention and they
interned all European civilians and imprisoned POWs, nearly
100,000 in all. All were made to toil like slaves, and within about
forty months more than half were dead from torture, murder, near-
starvation, disease, deliberate neglect or overwork. They laboured
and were maimed and killed on the infamous Burma–Siam railway,
or in the mines or factories of Japan or Manchuria. The Japanese
enjoyed humiliating whites, but unknown thousands of Chinese
were murdered in the East Indies, in massacres of often paranoiac
cruelty.

The Japanese Co-Prosperity Sphere was notably successful in
denying the enemies of Japan natural rubber, but not enough of
the product reached Japan itself because of the efficiency of Allied
(mostly American) submarines and aircraft. Throughout Japanese-
controlled Asia there were also food shortages, because many of the
islands had previously imported staple foodstuffs by sea from more
efficient growers; in 1943 and after, there were Japanese shipping
shortages due to the success of Allied submarines. This made
nonsense of most of the Japanese economic ambitions.

As with the Germans, no lasting benefit came from Japanese militarism, except for a *civitas* reborn in the Anglo-American mould. It could be said that the by-product of the Japanese occupation of the East Indies was decolonization. But there was also a resolution by the American victors and by liberated natives to seek colonial independence, an aim abetted by the new British Government formed by the Labour Party in 1945. The Party had preached anti-colonialism for nearly fifty years and now backed, in particular, immediate freedom for India. Throughout the East Indies, the Chinese had been the dominant commercial class before the Japanese invasion, and it was not difficult for the Japanese to aim racial anger at the Chinese who were both alien and privileged. But for Chinese and Europeans alike, 'privilege' was often short-hand for enterprise, aspiration and hard work. Later, perceived racial feeling led to Malaya and Singapore being regarded as separate future sovereign entities by the British Colonial Office, the Chinese dominant in Singapore, the Malays on the mainland. Under the enlightened despotism of Lee Kwan Yew, Singapore ultimately became a Chinese city-state, at least the equal of Hong Kong in economic vigour. The resounding success of the overseas Chinese in Singapore and Hong Kong is the greatest single hope for the future of China itself.

The so-called 'Emergency' in Malaya was, in some ways, worse than wartime. During the War, the Japanese had been the known enemy, most other races friends, potential or actual. Not all Malays had been collaborators but few Chinese had been fellow-travellers. In the post-war 'Emergency', no one knew which Chinese were actual, or potential, or part-time *enemies*; or could they be non-Communist or anti-Communist? The 'sleeper' system so cleverly exploited by Communists in Europe as well as in Malaya worked thus. By conviction or blackmail, or some other form of coercion, an individual would be recruited into the Party and he or she might remain undetected and undeclared until needed, months or even years later. He or she might then only be required to do one job – murder, blackmail or duress of some kind that might make an enemy's effectiveness or life impossible. Then the sleeper would escape into the Chinese community and hide until next required. At first, it might sometimes appear that the sleeper was the European's friend or associate, but in that case, if the sleeper were unsuccessful in the allotted task he or she would have to flee, or be unmasked. Loyal Chinese living near Communists were pressured into assisting them, helping the cause with 'presents' – sometimes obtained by extortion – of money, food, medicine, clothing or shelter. The duplicity and treachery of the situation made a rational response difficult to organize but rubber income was needed even more in the 1950s than it had been in the 1920s, Britain being even shorter of dollars. It was not until Field Marshal Sir Gerald Templar arrived in Malaya in October 1951 that the British could realistically hope that the Communists might be defeated. By 1954, the worst of the crisis was over in Malaya, but not elsewhere. In the same year, the French gave up in Indo-China, which ultimately led to US involvement in Vietnam, with known results: death, destruction, shame and evacuation twenty years later.

In late 1949, the Dutch had abandoned their sovereignty in the East Indies to a newly formed Indonesia, which started out in a state of much economic confusion, largely generated by the Communists. Indo-China exported little rubber for thirty years after the end of the War, except for a small tonnage to France. Thailand and Ceylon produced more than before the war, as did minor producers in Africa and South America. The political troubles in Malaya and Indonesia – the major producers of natural rubber – would have left the world very short if it had not been for synthetic, which was itself transformed by the development of a new synthetic, *isoprene*, in 1954.

But the new situation did not unduly handicap natural rubber. Synthetics in other products that had once been exclusively vegetable, from chewing-gum to vanilla flavouring, were often developed because of an absolute limitation in the supply of the natural product. In the case of other synthetics, such as quinine, it was hoped that the synthetic would be more effective as well as being a more secure source since it could be produced at home. With synthetic rubber, it was the price of crude oil that dictated the cost of the end-product, while there was no question of any real limitation of supply of natural latex, which over a given period only needs more acreage, or better trees, or new and more efficient techniques.

Crude oil is now in 2002 between eight and ten times the landed price of the same crude in 1954, in real terms, inflation ignored. This affects the competitive position of synthetic against the natural product, but there is another point to be made. The demand for rubber of all sorts has increased so greatly – perhaps by twenty times – since 1954 that there is no real question of a price war between the various synthetics and the natural product. In a condition of rising demand, every producer tends to be happy.

There was, however, one particular purpose for which natural rubber is supremely suitable – radial tyres. These tyres were originally very much a European phenomenon, but during the 1970s crude oil prices rose enough for mileage per gallon to become a US concern. This was largely because by the 1970s the United States had become a net importer of crude oil, and this economic truth has become more important every year.

So radial tyres spread across the Atlantic in a big way. Their use averted the need for the inner tube and they became especially important in the long-distance market, which was dominated by heavy trucks and inter-State buses. Using the new inter-State highways built in the 1950s, cars, trucks and buses progressively destroyed the railroads' *raison d'être*. Because radial tyres need a sidewall – but not a tread – made of a rubber with high hysteresis (i.e. a rubber with plenty of 'give'), and because economically produced synthetics have all been, up to the present, of a low-hysteresis type, natural rubber is essential for the sidewalls of radial tyres.

Because high-hysteresis sidewalls have to be of natural rubber, it is an economic fact that, with the current high price of crude oil, no petrochemical company is likely today to spend the billions of dollars necessary to develop a high-hysteresis synthetic. It is therefore not going too far to say that, without a high crude oil price and the resulting radial tyre, natural plantation rubber would have followed 'hunted' Brazilian rubber into limbo. It is ironic that the high price of fuel in the 1970s led the Americans to make such an issue of mileage per gallon and that this, in turn, led to the widespread adoption of the radial tyre in the United States. But the high price of crude oil makes it unlikely that anyone would invest the millions of dollars necessary to produce a high-hysteresis synthetic rubber, unless it was from some raw material other than petroleum. Some genius might yet find a means of developing a high-hysteresis rubber from maize or potatoes or bananas at a cost lower than any made from petroleum, but it is unlikely.[21]

So natural rubber is now probably more secure than it was just after the Second World War, before radial tyres and several oil crises and after a war that was largely won on synthetic. Without natural rubber, Malaya and Singapore would be much poorer than they are today and, probably, less happy. Indonesia might have found it impossible to survive as a sovereign state and would have broken into component islands.

So the material well-being of millions of people in the East Indies can be ascribed to OPEC raising the price of oil in the 1970s. To claim that Indonesia, Malaya and Singapore owe their prosperity to the seemingly-aggressive determination of Israel, which is what led OPEC to raise crude oil prices in 1973–4, might be going too far. But it is part of a sequence that historians have so far neglected. Meanwhile, rubber remains a commodity at least as valuable and vital as oil itself and, unlike oil, natural rubber can be grown instead of only being extracted.

It would, however, be a fairly safe bet that 'native rubber' would still be tapped, even if the East Indies were impoverished either by a low oil price or by a cleverly developed high-hysteresis synthetic rubber.

NOTES

1. It is only naïve moderns who are unable to imagine a world in which every drug, every food and fibre, every repeatable luxury had to come from the natural world, usually from the plant world.

2. Priestley did not stop with investigating oxygen; he 'discovered', or at least identified, ammonia, hydrogen chloride, nitrogen peroxide and sulphur dioxide. He was also, to the good of ordinary mankind, the first to 'carbonate' water with CO_2, to produce what we now call soda water. In Seltzer, in Hesse-Nassau, there is some naturally occurring sparkling water containing sodium chloride, calcium, and magnesium carbonate, and soda water was called 'seltzer' in Victorian times.

3. Before rubber, the only clothing which was virtually shower-proof was oilskin, made of thick cotton or linen sailcloth treated with linseed oil. This was expensive and not always effective and most people merely wore thick woollen clothes.

Condamine addressed the Royal Academy of France in 1745, and this academy – the precursor of the Académie Française, and the Paris equivalent of the London Royal Society – showed great interest in the possibilities of the substance. Copies of Condamine's address were sent all over France and to various French overseas possessions. One reached Cayenne, where François Fresneau had been building fortifications in this steamy outpost of France for fourteen years. Cayenne, in French Guyana, did not harbour any of the *Hevea* trees, but another species, *Mapa*, was found to yield latex. From this juice, mixed with that of a fig, Fresneau made a pair of boots.

4. There is an elegant *ad hominem* link between the political and industrial revolutions; between the political upheaval of America and France and the industrial transformation of England. Years before Independence, Thomas Jefferson's 'philosophic teacher' (or tame scientist) was one Dr William Small, who was introduced by Benjamin Franklin to Matthew Boulton. Dr Small became Matthew Boulton's friend and adviser, and introduced him to James Watt in 1767. In 1772, Roebuck, Watt's financial partner, got himself into financial difficulties, through matters having nothing to do with Watt, and two-thirds of James Watt's basic steam-engine patent was sold by Roebuck to Matthew Boulton. The firm of Boulton & Watt then became the greatest manufacturer of all sorts of metal machinery of every kind, from coin-stamping to the steam-engines which replaced human, animal and water power in textiles, mining and other industries.

James Watt also carried on a correspondence over many years with Lansdowne's friend, the chemist and dissident, James Priestley. Boulton and Watt in turn stumbled upon gas lighting as an alternative to candles when one of their employees, William Murdoch, who was installing pumping engines in Cornwall, suddenly devised a means of using coal-gas to light his house. Though coal-gas became a local convenience, and transmission by rigid metal pipe became possible over short distances, gas, like many other new products, had to wait for rubber to be of use in wide applications. It was not only a question of rubber pipe, so familiar in the laboratory of a century ago, but also of joints and adhesives and seals of all kinds; gas distribution was only one of a dozen great basic industries which were waiting for rubber at the beginning of the nineteenth century. Local gas-lighting was a novelty in 1800. By 1860, it was in general use all over the civilized world, to be superseded in turn by electricity. Neither would have been possible without rubber.

5. The first significant military use of the new weapon was at the Battle of Fleurus, during the early years of the French Revolutionary Wars, in 1794.

6. Vulnerability restricted the use of captive balloons before the American Civil War in the 1860s and it was only the invention of the Morse telegraph and the portable gas generator which had made their appearance at Richmond in 1862 possible. Sixty-four long-distance free balloons ascended from besieged Paris between September 1870 and February 1871. Only two disappeared. One reached Norway in fifteen hours. Captive 'observation' balloons became universal in European staff appraisals in the 1880s, and they were of use in their own right and prepared the military mind for other forms of aviation. Like his great opponent, Wellington, Napoleon preferred to work with proven, safe, predictable

weapons. Balloons went into the same category as Congreve Rockets and early shells invented by Major Shrapnel.

7. Textile garters for men and women had been in use since Roman times. Only with the advent of rubber did they become more effective. As with garters, so with self-supporting underclothes, shorts, knickers, skirts, long gloves, stockings, hats and caps. The *soutien-gorge* or brassière, that important essential in female form, that triumph of engineering and of appearance over reality, did not appear until the late 1800s.

How and when did sad, kinky men acquire the love of rubber, which was an apparently important minor sexual perversion by the time Freud was operating in the 1890s?

8. Goodyear's name is now known all over the world. Less well known is that he left six children who were brought up in intermittent poverty: the eldest son, Charles, was obsessed with the idea of making a shoe entirely by machine. He invented more effectively than his father, and remained as silent, as cautious and as prudent as his father was not. Charles Goodyear Jr founded the Goodyear Boot and Shoe Machinery Company, did all he set out to do, and died a rich man. He was even able to subsidize his younger brother, William Henry, an artistically inclined sibling who was the American exponent of the Gothic revival, doing in the United States what John Ruskin did in England.

9. The first commercially successful use of Goodyear's inventions was to vulcanize rubber thread so that it could be used to give cloth a ruffled appearance. This was called 'shirring' by the fashion trade of the day.

10. Brockendon, who proposed the term 'vulcanization' to Hancock, passes from history, dilettante to the last. Stephen Moulton, who had brought the samples from Goodyear and, via Brockendon, had taken them to Hancock, went on to start a great industrial rubber industry at Bradford-on-Avon in Wiltshire. Bradford, Wilts, which produced more woollen cloth than Bradford, Yorks, in 1830, became one of the early centres of the English rubber trade, thanks to Moulton. The former woollen mills became rubber factories, Moulton's name became a byword in the heavy industrial use of rubber, particularly in railways, and his partnership with George Spencer became famous as Spencer Moulton.

11. There was gum Arabic, used as an adhesive, emulsifier, or glaze in art and commerce for 2,000 years; or olibanum, a gum-resin from the *Boswellia* shrub growing in India and Arabia, and producing frankincense as an end-product, first exported to Egypt 3,000 or 4,000 years before 1851; there was shellac, derived from the trees of the genus *Ficus*, and used for thousands of years in India as a base for lacquers and polishes. *Ficus* also yielded a form of latex for rubber. All these vegetable products were derived from the sap of trees or shrubs and used, after treatment, in many pre-industrial countries. More modern analogues to the latex derived from the *Hevea* tree were turpentine, from various pine trees, Kauri gum from Kauri trees in New Zealand, used for varnish and to make linoleum, and dammar penak resin from Malaya, used in the paint and varnish industries and gathered in exactly the same way as latex from *Hevea* trees, by incision and 'bleeding'. All these products were exhibited at the Great Exhibition of 1851, as were edible products from sap like maple syrup. Before synthetic chemistry and

higher standards of life (and wages) in the Third World countries made vegetable origins no longer economic, there were more than 1,000 discrete vegetable raw materials of commerce, excluding the obvious ones like edible or drinkable material, timber, fibres, rubber and drugs. More than 3,000 vegetable drugs were used in England, some more effective than others. Most of the vegetable raw materials would have been represented at the Great Exhibition, in one form or another, if they had reached the commercial stage. Most of them would have been available on the London market.

12. In theory, solid tyres avoided punctures, but it was heat which destroyed their validity. At a speed of 20 mph, a tyre 36 inches in diameter revolves three times per second. This means that a part of the tyre is loaded (when at 6 o'clock) or completely unloaded (when at noon) three times per second, 180 times per minute. In the process, energy is absorbed and released and not all the energy in each revolution is released into the atmosphere. So part of the energy is converted to heat, which in turn is absorbed deep in the tyre. Dissipation of heat in rubber is difficult because rubber is a poor conductor, so heat generated within the mass of the solid tyre builds up. The heat is first absorbed by the central mass of rubber, which then liquefies, and the process continues until the liquid rubber hydrocarbons gasify. These gases are under pressure, so that ultimately they burst out, usually through the wall of the tyre, and puffs of smoke and smelt would be seen coming out of the disintegrating tyre. In extreme cases, the structure of the tyre broke down, and the tread was shed in large lumps. Long before this, the wise would have stopped. Even in the 1920s and 1930s, heavy trucks could be seen 'resting' outside some convenient hostelry while their solid tyres cooled down.

13. In all, a bicycle needed as high a quality of manufacture as a rifle, and in 1900, each cost about the same.

14. Berlin was the first city to convert to 220 volts AC to save local transmission costs, the first to install major turbines, the first to have ring mains necessitating the first oil-cooled switch gear; the lowest cost per installed kilowatt was in Berlin, as was the lowest conversion cost of coal into electricity.

In New York, it was the survival of the fittest that ruled, with consolidation only in the 1900s.

15. Ferranti developed alternators, switch-gear and cables of greater power and utility than had ever been seen before. By 1891, the whole of London could have been lit by electricity generated on the river, where coal could be brought in by ships, ashes removed, and water for cooling and condensing was freely available. But Ferranti was ahead of his time, and the Deptford power station was defeated by sectional, political interests because many local authorities had huge investments in gas undertakings throughout the London area, and these were threatened by electricity. The same local authorities had powers that could be used to block Ferranti's initiatives.

16. Above the *patrão*, there would be a number of other grades of exploiter, akin to modern drug-dealers. In the period between 1900 and the First World War, a famous Peruvian, Arana, 'controlled' 8 million acres of jungle lands. He operated first in Manaos, then in London. He employed eight to ten area 'chieftains'. Each of these in turn employed up to 200 'section chiefs', who in

turn each employed as many tappers as they could find. Each grade was in thrall to the next. The top people only got paid when they delivered rubber, as with the tappers and section chiefs. This was akin both to the feudal system and to the modern gangster dealing in drugs, and in each case there was a non-monetary trading arrangement.

It was all very wasteful. The 8 million acres which Arana controlled should have produced 16–25 million lb of smoked rubber, delivered to Manaos. In fact, in his best year Arana only exported 1.42 million lb. This was at less than 10 per cent efficiency, but not surprising. Systems based on brutality are rarely efficient, and the only really remarkable result of the efforts of Arana and other gangsters was the extent of the genocide which they consciously or unconsciously committed. In the years between 1900 and 1914, Brazilian tribes in the affected Amazon jungle are said to have lost more than 50 per cent of their population. In Peru and Bolivia, Ecuador and Colombia, the story was the same, and in the huge Amazon basin, bigger than most countries in the world, nearly as big as Australia, there were probably less than a tenth as many Amerindians in 1913 as there had been before the whites arrived in the sixteenth century.

17. In 1900, there were four sources of hunted 'wild rubber'. There was *Ficus elastica* from Assam, Burma and other places in Borneo and other East Indian islands. There was latex from Central and South American trees and bushes, from *manihot*, from *honcomia*, from *castilloa*, *sapium* and *parthenium*. From these original plants, nearly seventy different grades were produced, all of which entered commerce. African rubber came in 120 grades, from vines like *Landolphia*, and from trees in the *Fistuca* family. From Para in Brazil came the only really top grade rubber, from *Hevea*, half the world's supply, against which all other rubber was judged.

In 1910, the relative volumes were: wild rubber 80 per cent, plantation rubber 12 per cent. In 1913, the figures were wild rubber 40 per cent, plantation rubber 60 per cent.

18. Rubber was much less important to Brazil before 1850 than three other commodities, sugar, coffee and cocoa. The last was so important to Para that the city had been on a 'chocolate standard' since about 1770.

19. See page 153.

20. As an illustration of the ups-and-downs of those times, there is the story of one famous Namyang (Overseas Chinese), Lim Hoy Lan, fluent in English, the son of illiterate parents. He arrived in Singapore at the age of twenty, with a few pounds sterling in his pocket. Two years later, in the spring of 1920, he was paper-rich. Raw rubber was then over 60 US cents per lb. In September, the price had fallen to one-quarter of this – 15 cents. Thousands were ruined, among them Lim Hoy Lan. He was young enough to laugh, however: he had enough cash to buy a steerage passage and he had only $24 Hong Kong in his pocket when he arrived on that island. He went on to make several fortunes and, with Peter Tsui, he founded the Wah Yan College, one of Hong Kong's best.

21. Many synthetic polymers have been developed in the last fifty years for industrial purposes, in electrical fittings or in footwear, for example. None has high-hysteresis characteristics at the commercially correct price.

TOBACCO

More Than
a Smoke

I

There are sixty-six species of the tobacco genus, *Nicotiana*, which is named after Jean Nicot, French Ambassador to the Portuguese Court at Lisbon. In 1559, Nicot was given seeds of a Brazilian cultivar, later identified, classified and named *Nicotiana tabacum*. This soon became the only really important species for smoking, sniffing or chewing. Of the sixty-five other species, except for those grown for their handsome flowers, most are to be avoided. Some are so high in toxic alkaloids as to be poisonous, some varieties being those from which insecticide and even, at one time, rat poison was made, an infusion from which was the means of suspected murder in a few cases.

Although Jean Nicot soon convinced the French Court of the joys of tobacco, its use by non-aristocratic French trailed somewhat behind its use by the English, partly because tobacco became a Royal Monopoly in France. Tobacco was a top-down fashion in France, while the freer English, especially humble English sailors, soon imported and used tobacco to chew as a solace and a mild intoxicant. By 1580 tobacco was a garden crop, not only in England but also in Portugal, Spain, Italy, Switzerland and, to a lesser extent, in France. By 1620 tobacco was grown in most countries in Europe, and had been taken by Western Europeans to Russia, Turkey, the East Indies, Japan, China, Korea, India and West Africa. Ironically, the British-American mainland, so important in tobacco history, was a late starter. But by 1650 there were few countries with which Europeans traded which did not grow tobacco, either for use at home or for export. The way that tobacco spread so quickly across the known world and was used by so many so soon is a tribute to its obvious attractions, which included creating wealth. The question of addiction did not significantly arise for at least 300 years.

The most important country not yet 'discovered' by Europeans – Australia – was possessed of a native species of *Nicotiana*, which Aborigines used in much the same ways as did indigenous Amerindians. The Australian species, *N. bethamania*, did not commend itself to British smokers who arrived with the First Fleet in 1788 and no one today would use *N. bethamania* except for plant-breeding, to develop a new hybrid. Some Aborigines in Australia showed the white invaders that smoking was a form of near-sacred herbal therapy and they enjoyed the same faith in the remedial qualities of the weed as did various Amerindians who were found smoking, sniffing or chewing by early Italian, Spanish and Portuguese arrivals in the New World two centuries before. In the Renaissance world where all plants were thought to be valuable for food, poison or medicine, and before any logical form of botanical or chemical analysis, tobacco was held by some Europeans to be a 'cure' for New World diseases. These included syphilis, brought back from the West Indies soon after Columbus's first voyage. In this case, the peculiar notion that tobacco was a cure was soon found wanting.[1]

No wild example of *N. tabacum* has ever been found by a botanist, so it has to be assumed that Amerindians domesticated the species and helped it to hybridize, which *N. tabacum* does with great ease. Alternatively, other early wild species may have interbred in such a way that they became self-sterile, an unintended fate too easily achieved, even today, by impatient plant breeders.

The botanical family to which inedible tobacco belongs, the *Solanaceae*, includes four food plants native to the New World and of course unknown in Europe in pre-Columban times. These were capsicums (sweet peppers), only really widely used in Western European cuisine since the Second World War. Then there were tomatoes, without which it is difficult to imagine Italian pizza or pasta. Although many of the eggplants (or aubergines) are East

Indian in origin, many of the more useful and attractive cultivars came from South America, again of course after Columbus. Far more economically and historically important is the potato, whose original home was the Tropical Andes, but which is now an important food grown from the near-Arctic in the north to the far south of the Americas.[2]

*

Tobacco of a sort can be produced almost anywhere except in the Arctic and Antarctic regions, but seeds were originally brought by early Iberian explorers from Tropical America. After the great voyager Christopher Columbus was told that smoking tobacco would prove to be a cure for various diseases, a common enough claim for all sorts of plants at the time, the argument was much used in the early days, wittingly or not, like the shame-faced argument today that a smoker only smokes to help lose weight, since nicotine suppresses the appetite for food.

In the Renaissance, Europeans still believed in the system of the four bodily humours – blood, phlegm, choler and melancholy – developed by the second-century Greco-Roman physician Galen. Everything consumed – food, drink, herbs or spices – had to correct or intensify an individual humour or a combination of humours and it was the aim of the healthy consumer to maintain a balance of the four. It is easy to see what spices like ginger or pepper might have been said to do, but many of the new American imports were more difficult to fit into the Galenic mould. Although sweet peppers were obviously hot, like ginger, tobacco did not seem to fit any humour with any conviction. But some authorities thought that tobacco smoke expelled phlegm and warmed and dried the body, while chewed tobacco had the reverse effect. What is incontrovertible is that tobacco tends to suppress appetite, which was very attractive in a world that often went hungry because there were usually too many people competing for too little food.

It was more than a century after Columbus's first voyage before tobacco was accepted simply as a recreational herb regularly used for pleasure, but by the mid-1600s the tobacco trade had become economically important and had made many men rich. Few countries were then using more tobacco per head than did the people of England. But the Dutch in their heyday at the time of Rembrandt, in 1650, were said to be using twice as much per head as the English; the claim is difficult to check, however, since the re-export tonnage from Holland is less easy to determine than tobacco exports shipped from England.

It was the Stuart King James I who had determined that all tobacco from Virginia consigned to any place other than England should pass through designated English ports, at first only London. This was to ensure that the tobacco trade remained under Crown control, direct export from the Colonies to non-English territories being seen as potentially subversive to a would-be despotic monarch and his bureaucrats. The demand that exports to every country should pass through British ports was resented in the American Colonies and by the 1730s had become a festering sore that helped fuel revolutionary spirit in the 1760s.[3]

James (I of England, VI of Scotland), was not only instrumental in creating a substantial grievance about the restraint of trade by the Crown. He also, conversely, favoured the new colony of Virginia whose future wealth was an unintended consequence of Royal Stuart tobacco policies.

It would be an irony that James's grandson, Charles II, would have enjoyed, if he had known that Stuart tobacco trading policies would lead to a class of American tobacco landowners, generally more enlightened than their English counterparts, who would inspire and successfully lead the War of Independence. By 1776, at the time of the Declaration of Independence, a century after the reign of Charles II, a quarter of the signatories were involved in the tobacco trade. Before upland cotton, tobacco was the field crop that yielded the greatest wealth and it was tobacco that produced plantation owners like Washington, Jefferson, Madison and Monroe.[4]

In 1607, at the time of the first Virginian settlement at Jamestown, it is computed that tobacco was already used by a quarter of the English male population more than three times a week, probably chewed or burnt in a pipe – perhaps more often than not in a communal pipe – as in an alehouse. Enough tobacco (not from Virginia) had been imported in 1607 to supply English males with 200g (7oz) each year. This meant that the quarter of males who smoked used 800g (less than 2lb) of tobacco each year, enough to fill about two small pipefuls each day, equivalent to about four modern filter-tipped cigarettes. Not much by modern standards, and never enough to point to any kind of addiction. But tobacco in 1607 came from non-English sources and cost the smoker a relatively very large sum of money – ten times as much as it did fifty years later. And tobacco imports cost the English economy large quantities of gold, virtually the only way then of settling a deficit in the balance of payments.

James I had issued his 'Counterblast' against tobacco as early as the second year of his reign, 1604, but economic necessity drove him to approve its growth in English, not foreign, colonies. This was because the most desirable species, *N. tabacum,* came originally from what is now Latin America, and seeds or plants had to be transferred to British-American colonies if acceptable tobacco was to be grown in the Colonies for the United Kingdom market.[5]

The American Colonies, rather than tobacco plantations at home, were favoured for three reasons. First, imported commodities were easier to tax and to control than those grown at home. Second, if Colonies were to be established, they must produce something the Home Country wanted or needed and which could be traded; colonial produce would be exchanged for goods and services from England and gold from foreign countries. Third, there was no point in establishing Colonies unless they were to be a source of strength, and not be or become a liability. This pointed to trade in commodities of value. Because foreign trade transactions had to be settled in gold or silver, almost everyone who went to the New World originally looked for precious metals. Only when such searches proved vain in most areas did the agricultural value of the new lands become important.

From the first, most English Colonies were established for commercial reasons and within the three criteria above. It was only in later times, when Franco-British rivalry became intense, that settlements were established for strategic reasons, like Gibraltar in 1713. Later still, Imperial settlements would include coaling stations for the steam navy or unrewarding real estate in Africa during the nineteenth-century scramble. But seventeenth-century Colonies usually grew primary products that could not be grown at home.[6]

Besides tobacco, the main colonial crops that supported English settlement in the 1600s and created wealth and trade were indigo and sugar. But in the first, early days, before 1650, tobacco was the most profitable crop all over the English Americas, with the incomes of individuals higher – more than double – those at home. In many places tobacco provided only the original exports. In Barbados and other English Caribbean islands, sugar, ginger or indigo became economically more important than tobacco by the second half of the 1600s. This change coincided with the substitution of black African slaves for white European indentured servants. Before 1660–70, white indentured servants were far more important in the English Caribbean and mainland Colonies than the few black slaves then in the Americas.

Then there was the question of tobacco quantity and quality, which favoured the Chesapeake Colonies. By 1680, Virginia and Maryland were shipping more than 20 million pounds (about 9,000 metric tonnes) of tobacco, more than all the rest of the Americas put together. Chesapeake tobacco also commanded higher prices in Europe than did most tobacco imports from the Caribbean.

In this chapter it is not proposed to treat at length the pre-Columban use of tobacco by Native North Americans. Unlike Amerindians in the Caribbean or what is now South America, the natives of the Chesapeake region used not the bland *N. tabacum*, but other species higher in nicotine content and including many other alkaloids. Such tobacco use was likely to produce hallucinations and the plants were thought to have supernatural powers and even to spring from transcendental origins. These theories have been attractive to those who in our own time have sought to justify drug-generated escapism by the search for historic precedent – to build up intellectual respectability. Suffice to say that hallucination and accompanying shamanism are unlikely to be fuelled by *N. tabacum*, since there are few alkaloids in the species to produce hallucination. But the Amerindians whom the early English colonists first met at Jamestown in 1607 did not smoke the relatively mild *N. tabacum* and the very high death rate that the settlers experienced may have been due to their using a local species of tobacco that had a very high nicotine content and which, overnight, could have proved to be a killer. Conventionally, the low survival rate of sixty men out of 500 during the winter of 1610–11 has been ascribed to starvation. But perhaps it was in fact due to an excessive use of the local species of tobacco, *N. rusticum* – coarse, short-leafed, high in alkaloids other than nicotine, with a notably irritating smoke and thought by some to be virtually poisonous. It was John Rolfe, the husband of Pocohontas, who first pointed the way to the future in 1612. In that year, he planted *N. tabacum* and harvested the species for the first time in Virginia. The seed was brought from Trinidad, and in those days it was called Oronoco tobacco.

Optimum cultivation methods had to be learned from local Amerindians, whatever seed was employed. Techniques were quickly developed that remained in use for over three centuries. Tobacco seed itself is tiny, smaller than onion seed, so small as to demand special treatment. There are nearly 5 million tobacco seeds to the pound, more than in a pound of onion seed, the smallest seed in general use in England in the 1600s.

A special seed tray had to be used, and was more effective than any native method of sowing as a field crop. A seed tray allows tiny seeds to germinate without the competition of weed seedlings which would have smothered the small tobacco plants. Any diseases and pests can be seen and affected plants discarded before the pests contaminate a whole field. The soil in the seed tray can be simply sterilized and generously prepared with plant foods fit for germinating seeds. There is also the point that if the season for growing tobacco is limited by the length of the wholly frost-free period between late spring and early autumn, the vigour of seedling growth in the tray is an important factor in expediting the whole process. There is also much evidence that the mere act of trans-planting seedlings from tray to field accelerates and physiologically benefits rapid growth.

Sown in trays, or in specially prepared frost-free beds, before the middle of January, seedlings were transplanted into the open field in April. This was a near-crucial operation and needed much skilled labour. It was done when the ground was wet yet friable, with plant root systems fully developed but with the actual plants still easy to transplant. During transplantation it was vital to move quickly as tobacco plants suffer if allowed to dry out, even for a short time. Fields into which plants were to be transplanted were nearly as carefully prepared as the seed trays or beds in which they germinated. Several thousand plants per acre was the aim, but more when fertility was high enough.

Amerindians, having no draught animals and no ploughs, planted their seedlings on little hummocks or hillocks of clean prepared soil about 6 inches high. These humps were arranged in long lines, rather less than 3 feet – an average non-marching male pace – apart, and these would produce rather fewer than 5,000 plants to the acre, which is about 70 yards × 70 yards. Planting on hillocks or hummocks would have several advantages, with soil properly prepared and relatively weed-free, and naturally good drainage. Copied by the white man, these practices were derived from local Amerindians in the absence, universal in Virginia in those early days, of draught animals.

Once transplanted, tobacco plants grew leaves coincidentally with the leaves of weeds that surrounded each plant. Skilled hand-weeding became essential, but later weedings harmonized with two other requirements, to top the plant to prevent it flowering and to encourage maximum transfer of nutrients to the leaves, the whole object of growing tobacco in the first place. But as soon as they were topped, tobacco plants naturally threw out suckers which then had to be removed.

Weeding was a continuous preoccupation until the arrival of chemical control in the middle of the twentieth century. Timing for topping was critical and skilled judgement was needed in the case of each individual plant. Early topping tended to reduce yield by as much as 10 per cent but late topping reduced leaf quality as well as yield. In the 1600s, the average yield of leaves was probably about 500 lb per acre. It is calculated that early or late topping reduced yields by as much as 15 lb (3 per cent of 500 lb) per acre for every day that the cultivator, or his workers, missed the optimum time for topping.

Tobacco was a crop which, after planting, offered few opportunities for growers to sit in the sun and watch the natural increase.

This idle pleasure could be enjoyed with corn (maize), wheat or potatoes, but never with tobacco. Having finished the last – perhaps the fifth or sixth – hand-weeding of the crop, the next vital determination had to be made in September, when the decision as to when to harvest – to cut the leaves off the plant – had to be guessed at, judged or resolved.

Early harvesting produced leaves that could never be suitably cured and the resulting damp mess was worthless in the market-place. Late cutting, on the other hand, risked a frost that would do even more damage than cutting too early. Tobacco farmers were fraught, anxious and stressed out in the early autumn unless, of course, they had a lot of knowledge of their own local micro-climate. When ripe on the plant, tobacco leaves are greyish, no longer green in colour, a visual hint that sap no longer energizes the growing extremities. Ripe tobacco leaves thicken and tend to be less sticky and oily to the touch. When pressed between thumb and forefinger, the leaf must feel as though it will crack with ease, indicating that growth has ended and that the leaf is ripe and drying out. This is all the stuff of local knowledge, instinctive plant husbandry and good field management, and is difficult to teach in college and has to be learnt over the years.

The decision to cut having been made, and the calculation established that all the leaves could be harvested before the arrival of a frost to threaten the year's work, there was another factor to be taken into account. Leaves picked green – or, more properly, grey-green – were unmarketable until they had been cured in a barn for up to three months. But the field could not just be harvested at once, like a field of grain. Harvesting a tobacco crop by hand was much more akin to gathering strawberries, each fruit or leaf picked only when ready and ripe. Each tobacco plant, each leaf even, had to be judged to be fit to harvest and picked at the right stage – a skill only acquired through experience. It is worth mentioning that tobacco can grow more than 7 feet high and the judgement as to fitness might have to be made by a worker much shorter than the crop itself. It is notable that English farm workers in the Americas had to deal for the first time with crops as high or higher than a man – tobacco, maize (corn) and, in warmer climates, sugar-cane. This must have been an odd experience, even for seasoned and experienced farm workers who had emigrated as indentured servants.

Tobacco leaves then had to be hung up in bunches in a purpose-built barn. The barn had to be aligned to attract the natural flow of air in relation to the prevailing wind, taking into account the configuration of the ground and the movement of local breezes. The nature of the barn and its ability to dry or condition tobacco leaves dictated, in part, the condition of the leaves picked in the field. But the process in the barn was itself critical and wholly dependent not only on natural airflow but also on ambient relative humidity. In a damp, foggy, windless autumn, the drying process would be many times slower than in an autumn distinguished by bright, sunny days and the kind of winds associated with these conditions, and a relative humidity lower (and more helpful) than normal. If too dry, leaves would disintegrate as dust, while if too damp they would rot. When judged to be cured, the leaves were packed for sale, first stripped from the stalks and then packed into hogsheads. Since ship-owners charged per hogshead instead of by weight, the barrels were packed as tightly as possible, so that a hogshead might weigh as much as 1200–1400 lb or more, the product of more than 2 acres in the 1600s. Once the cycle of harvesting and packing for export was completed, it was already past the proper time to sow the next year's crop. Tobacco was a harder task-master than any crop previously grown by Europeans.

Tobacco was a unique crop for the English in Virginia in three other ways, and these were interlinked. First, tobacco is a gross feeder and needs nitrogen, potash and calcium in large quantities. The actual amounts needed per acre were of course unknown until the twentieth century and in the 1600s rotation of crops was only recognized in the crude form that had existed in the Middle Ages. Wheat, then barley, was grown on good ground which was then fallowed and left to restore itself for a year, while cattle and sheep from the village roamed it at will, leaving their invigorating manure in a random manner. One crop of grain was grown on less good land, followed by a year of fallow; in poor districts the crops would be rye, barley or oats, not wheat. (Many people in districts of poor land in the Middle Ages would live and die without ever having eaten risen wheaten bread.) On really poor land, there could be only one crop of grain in three years. In the Virginian soil, without housed cattle to produce manure, English settlers grew one or even two crops of tobacco, followed by as much as twenty years of fallow. This pattern was unique to tobacco land.[7]

In heavier land, this regime worked for two crops, but in lighter soils more than one tobacco crop often destroyed the land's fertility beyond restoration. By the time of the Revolution, much lighter land had been deserted and had reverted to scrub or second-rate woodland. Former Tidewater owners moved on; this happened again and again in Virginia, farmers pushing the frontier further west as they moved.

There was too much easily reached, freely available land in America for men to bother with what was worn out, which consequently became barren, weed-ridden, unprofitable or otherwise unsatisfactory. A huge, empty country was responsible for creating a footloose white population. Rejected and worn-out land was as common before 1900 as rejected wives became in the twentieth century.[8]

Because each acre needed up to twenty years for the restoration of its fertility after growing tobacco, and because land was also needed to grow food, timber for fuel and barns, fences and for making hogsheads, the requirement per active worker was a large one and dictated a new ideology for farmers and landowners used to English conditions. It introduced the idea of production-per-man-hour (PMH) in place of the traditional output-per-acre as a way of measuring efficiency, an idea that would endure in American conditions. This was partly, but not wholly, because land was cheap, available and 'empty', and partly because labour was often in short supply. It was more than 300 years before the same management concepts became widespread in England, where farmers had to wait for mechanization to make PMH a reality.

Second, unlike many other forms of crop husbandry in the 1600s, tobacco culture needed considerable seasonal judgement in the humblest worker. A sense of appraisal was needed in at least five or six stages each year. This meant that labour could not be merely acquired and put to work. So when slavery was adopted in Virginia after the 1680s, while men off a ship could be put to work further south in fields of rice or indigo or sugar, field workers had to have an instinct about tobacco, or to have grown up as children in the tobacco fields. This tended to mean better relationships between masters and slaves in Virginia and other tobacco areas than elsewhere in the South.

Third, if land was not the real economic limitation, a financial and economic pattern was established. Agricultural land – in contrast to urban real estate where location was vital – was worth little more than the value of what was to be found in it – minerals, timber, roads, fences, drains, buildings and so forth. This is still much truer than not in the United States and far more so than anywhere in Western Europe, for obvious reasons.

By 1700 a social pattern was established in the tobacco lands. A few landowner-growers were already becoming richer and would adopt the style and manners of plantation owners within a short time. In 1700, European indentured and ex–indentured servants very comfortably outnumbered African slaves in the Chesapeake Colonies. But by the year of the Declaration of Independence – 1776 – there were virtually no white indentured servants left and the black population of Virginia was over 40 per cent – about 230,000 out of 550,000. Most tobacco, by tonnage, would be grown in Virginia itself after 1750, not by smallholders but by men like Carter, Lee, Byrd, Washington and Jefferson – that is, planters, landed gentry and slave-owners, not working farmers; most of the time, they wore fine clothes too grand for manual work and were unlikely to go into the tobacco fields except on horseback. Their humbler neighbours had already moved West.

II

The wealth of the tobacco trade in Virginia was important in both the Revolution and in the war between the States. In 1763, the French and Indian War, called the Seven Years' War in the United Kingdom, came to an end. It was the fourth of the conflicts between France and England since 1689, in all of which the Americas had been involved to a greater or lesser extent. In the peace terms of 1763, the French conceded their colony of Quebec and six West Indian sugar islands, as well as their territorial position in India. Two islands, Guadeloupe and Martinique, were restored to France, as were fishing rights on the Grand Banks, while other territories were returned to France or to her ally, Spain. Many politicians in London were critical of the generosity of the settlement as far as concessions to France and Spain were concerned. But the majority of Continental diplomats feared the great increase in British Imperial power and it later proved difficult to prevent most European countries from supporting the American Colonies in the War of Independence. Their pro-American stance was provoked much more by anti-British sentiment than by any commitment to Life, Liberty or the Pursuit of Happiness. As usual, the exercise of the balance of power was driven more by envy and fear than by altruism. After 1763, the United Kingdom was top dog and became the common enemy.

It is clear that critical sentiments about the British Government lay not far beneath the surface in the attitudes of the American colonists after the Peace in 1763. The greatest grievances concerned trade and taxation, and grumbles about the new enhanced Imperial power of the Government in London were less important.

The ultimate independence of the Colonies was due to the facts of geography and this should not be underestimated, however appealing the idea that it was entirely due to noble altruism. In 1763, westbound ships from Bristol, England, took seven to nine weeks, plus several days extra at least from London, to cross the 3,000-plus miles of ocean to New York, more to Chesapeake Bay. With luck, and a prevailing wind and currents, eastbound passages were easier, only four to five or six weeks, sometimes less than a month. But a despatch, and a considered reply, would take at least thirteen weeks to cross both ways.

For these obvious reasons, correspondence from the London Government tended to be ignored or considered out of date, and men on the spot often thought they should have been left alone to operate as they saw fit. But that is exactly what did not happen after 1763. Pitt the Elder had fought a very successful war by choosing good men and allowing them to get on with the job. But his less able successors thought they could govern the Colonies by exerting a degree of control that geographical realities made almost impossible to achieve.

Abetting the mediocre thinking in London were the deeply held convictions of the young King George III, 25 years old in 1763. He was relatively uneducated, especially in history, but – a very dangerous combination – he had a certain natural ability and an immense capacity for hard work. He regarded the British Empire as a royal fief to be preserved, frozen in some time that never was; the colonists, especially those in America, were to be submissive to the Crown. The Crown in Parliament was a concept derived from 1689, meaning that the King and his ministers should plan and propose, while Crown initiatives were endorsed by Parliament, whose constitutional duty it was to scrutinize, audit and of course control the finances of the Government. George III was intellectually unable to separate his principles from his prejudices. These were to rule much more forcefully than the previous two Royal Georges, they having been more Hanoverian-German, he more English, and these views were held by a largely unlettered young man, not noted for robust common sense. In fact his views on the King's place in Government were Continental, not English.

George's ministers, even his one-time tutor, Bute, were unable or unwilling to help the King see the situation in a more realistic perspective, and time was wasted because of the state of the King's literary abilities. In short, he found it difficult to absorb a reasoned written argument and harder to compose one. The King's character was a random piece of luck – good or bad, depending on how you looked at it – and this Royal hazard coincided with a general lack of quality in the King's ministers, since he naturally tended to prefer those who agreed with him.

Any hope that conciliation would triumph over the forces of coercion was blighted by an inability to accept that geographical rules applied to inter-colonial affairs as much as to the existence of the broad Atlantic. An innocence about the time needed to cross the ocean was compounded by topographical ignorance about the Colonies, the future thirteen United States. These extended 1,200 miles from North to South, and from 100 miles from the ocean in the East to the top of the mountains in the West, which largely formed the frontier in 1763. Land transport was slow, uncomfortable and expensive, and most inter-colonial trade, passengers included, went by water. American historians are apt to paint a picture of like-minded colonists with the same modest, reasonable aims, part-selfish and economic, part-noble and altruistic. In fact, of course, the interests of people in various Colonies differed as much as the nature of the Colonies themselves.

The original settlers had left England for disparate reasons; New England was largely settled by Puritans before 1660, especially Massachusetts, which was in 1763 a most populous Colony and led popular opinion in New England as a whole. But Rhode Island and Connecticut had liberal charters and elected their own governors instead of having them imposed upon them from London. Coincidentally or not, this made these two small Colonies more loyal to the Crown than were their larger New England neighbours, Massachusetts and New Hampshire. While Virginia had been largely settled by Anglicans and Maryland by Catholics, there were descendants of the original Dutch Calvinists on the Hudson River, a few Germans settled in New Jersey and Pennsylvania and some Protestant Huguenots in the two Carolinas. New York, not yet a multi racial city, was devoted to commercial concerns, as was Philadelphia, but there was also a residual Quaker sense of the ethical in the City of Brotherly Love. Overall, the English, Scots and Welsh and a few Scots-Irish Ulster Protestants represented 90 per cent of the near-2 million Americans in 1763. The white population doubled roughly every twenty-five to thirty years, to 2.5 million by 1776, plus an unknown number of black slaves. Before 1846, there were few Irish Catholics in America and virtually no Italians, Russians, Poles or Jews.

These mostly English men and women had English views about liberty; they were, for example, descended from the only country in the world that had had *habeas corpus* in some form for more than four centuries. But the English had also been through a Civil War, a radical revolution that deposed and executed a king, his son's Restoration and a great Settlement after a later, shorter period of would-be royal autocracy. The Settlement of 1689, the Glorious

Revolution, secured certain rights to individuals and to Parliament. These rights also institutionalized the doctrine of 'No taxation without representation'.

But grandees in London in 1763 were ignorant of how much freedom even the humble colonist enjoyed compared with his equivalent at home. This led to misunderstandings. Puritans in New England had not, as had many English Puritans, abandoned their radical faith for a more mellow creed; descendants of indentured servants in Virginia had not forgotten their forefathers' burdens; slave-owners in the South, like the same sort of proprietors in the West Indies, thought that those at home did not understand them. In every Colony there were many (perhaps a majority) who just wanted to live out their lives and improve their lot with as little hassle as possible, above all with a low level of intervention from the London Government. New taxes imposed by London would often provide the pretext for discontent, but it was probably those regulations that gave metropolitan merchants the lion's share of trading profits from British-American commerce that generated as much dissatisfaction.[9]

All traffic had to be in British ships, even, unrealistically, trade between the Colonies. All American exports to every country except the United Kingdom had to pass through designated ports in England or Scotland. There were import taxes on some American-produced commodities but, when the commodity was re-exported, the tax was remitted to exporters from the United Kingdom by means of the 'drawback' system. When this was applied to tobacco, which already produced one-quarter of all Customs' revenues in the 1660s, and more later, the revenue drawback hardly ever reached the American exporters.

In fact, re-exports largely benefited English merchants, and exporters found it easy to modify prices paid to importers so that most of the benefit of remitted duty stayed with the exporter from Britain and never reached the American farmer who had grown the crop. After 1758, when the duty was raised to 8 d per pound, to be compared with between 7 d and 1 s payable to the primary producer, according to quality, tobacco drawback on re-exports obviously became a vital component in the trade. These complex regulations did little to raise prices for growers in the Chesapeake area, but sometimes smuggling rose to 50 per cent of the trade between 1758 and 1763. By the latter date, other primary producers in the Colonies had other grievances against the insensitive way in which the metropolitan Government enriched English or Scottish traders at the expense of the nominally equal colonials. But the chief perception that excited American tobacco growers and traders was that the system was heavily rigged against them in favour of those at home. Hogsheads exported directly from the Chesapeake to the Continent would, it was said, remit half as much again to the Chesapeake growers as the same hogsheads that had to pass through the crime-ridden, sticky-fingered denizens of the Port of London. The same argument was valid for dried fish from Connecticut, timber from Massachusetts or iron from New Jersey, and trade with the British Caribbean was so encumbered by regulations that smuggling was much to be preferred to more formal commercial methods.

During the French and Indian War, many merchants in the Colonies found it convenient to ignore the Molasses Act of 1733 and trade illegally with French enemies in the West Indies. French planters, unable to export their produce to France because of the British blockade, sold it cheaply to American merchants. These in turn exported rum and refined sugar (or traded with the Indians) more cheaply than could those who had to pay both duty and an enhanced West Indian wartime raw-material price. This profitable ploy came to an end with the Peace in 1763 and some merchants considered the coming of peace as yet another grievance.

To the London Government, the loss of Customs revenue through smuggling had to be added to the cost of the war and the cost of the future defence of the Colonies. The British Exchequer was out of pocket and determined to make good the deficiency. As the British at home were more heavily taxed than the British in America, it seemed only fair that British-Americans should contribute to their own defence. But there was no real consultation within the Cabinet or with any representative British-Americans. If it was necessary to station, say, four to ten battalions in North America, then an equivalent of £4–£10 today million would have been needed to pay for them. This was then equal to about £10,000 per battalion for food, clothes and shoes and flint and powder. Artillery and cavalry would have cost more per man and per unit.

At the time, the economically active inhabitants of the thirteen Colonies would probably have numbered only 200,000 out of 2.5 million, since families were much larger and few, if any, women were economically active. Each battalion would have therefore cost each of the 200,000 1s, or one-twentieth of a pound. This would be equal to £5 in today's money, a bearable levy if paid by those most able to pay, but 1s in 1763 was equal to a day's cash wages for a labourer in an economy notoriously short of coin.[10]

Raising the money would be a major problem. There was no Income Tax at the time and Land Tax, the chief impost at home, was unsuited to colonial conditions. This left the Customs and Excise, and there were several, now famous, efforts by London Treasury bureaucrats – Stamp Tax and the Revenue Act, which included the infamous tea duties. These efforts ended in failure, tea duties because of the Boston Tea Party and its sequels. Nor, in the next twenty years, did any levy raise one-quarter of the cost of collection, a sorry comment on the lack of analysis by those in power in London. In the end, the War of Independence cost the Royal Government not only the thirteen Colonies but also more cash than the triumphal Seven Years' War. Those who start great conflicts little know their true money cost, let alone the cost in lives before the account is closed by victory or defeat.

None of the proposed burdens became a need perceived by colonial taxpayers; the French had been defeated and could be regarded, if anyone so wished, as no longer any kind of great threat. It was a tradition for the British – at home and abroad – to regard a defeated enemy as an authentic newly reformed character. Any belief in revenge as an important emotion held by the defeated was suspended. This was so even though there was much evidence to the contrary. Previous enemies – Spain, Holland and, later, France again, and then Germany – have all sought revenge. Disarmament after victory is an example of the English power of wishful thinking or of determined self-delusion.

There was another, unconnected, folly committed by the London Government because no one had the knowledge or the imagination to foresee the consequences, intended or unintended. This was to deny British-Americans any right to settle beyond the so-called Proclamation Line of October 1763. This effectively limited British-American settlement to the eastern flank of the Appalachians. The frontier between colonists and Native Indians would now be the watershed, preventing any westward expansion by those too enterprising (or too footloose) to remain in the more crowded East. The increase in population – doubling every generation – meant a permanent hunger for land, which was then the most important source of true wealth.

After this new frontier was imposed upon them, some Americans adopted the belief that it was a pretext for the presence of an otherwise unnecessary British army. The London view was that Indian chiefs and their followers and allies represented native power which could be friendly or not. Inevitably, white and red Americans would compete for the same land, if they were allowed to, so it would be wise to prevent any clash. How wrong the London-British were is shown by subsequent reality; nine States between the mountains and the Mississippi joined the Union before Maine did in 1820. But the British at home consistently underestimated the motivation, competence and strength of the colonists. Each State had to have a population of more than 50,000 to gain admission to the Union. There were over a million in the nine new States by 1820.

*

In 1763 George Washington was a landed (but not then rich) gentleman of thirty-one who had retired as a colonel in the Virginia Militia five years previously; he came back to his estate in 1758, where tobacco was the cash-crop. But he complained on his return from the French and Indian War that he could not get the high price for his tobacco that Tidewater neighbours had previously found possible. So in 1759–60 he turned his land over to wheat, a crop that needed far less skill and less adept labour and management. Profits from wheat husbandry were also lower, between a quarter and a tenth of those from successful tobacco crops.

Washington, the first President of the new nation in 1789, was the hero of the young Republic, responsible, as Commander-

in-Chief, for those American successes that could not be charged
to British incompetence. He was also, by then, one of the richer
men in the United States, his wealth largely acquired through
inheritance. But forty years before, as a poor boy of seventeen,
without prospects, he had taken up surveying as well as the practice
of hunting with firearms and dogs. The hunting led him to become
a successful soldier and commanding officer only half a dozen years
later. But the surveying may have been as important for the history
of America.

As early as 1740, by which time black slaves outnumbered
Tidewater whites, it was obvious to all that tobacco could only be
grown profitably if growers robbed the soil and moved on. In the
Tidewater, a leisured aristocracy lived on large estates like their
equivalents at home, built gracious houses, ordered furnishings
from London or had copies made in America, even ordered their
clothes from London, while their children were often educated
across the Atlantic. The difference between Virginia and England
was that in Virginia, there was an alien, subjugated slave workforce.
These slaves were inevitably less efficient (because of their lack of
incentives) than the poor whites who increasingly went inland to
the then West. The Tidewater rich owned enough land to give the
ground that had grown tobacco respite to grow another crop in
many years' time. (There were neither fertilizers to be found at that
date nor any known rotation of crops to restore the soil.) For the
poorer whites, it was a question of mining the stored fertility of
the land and then moving on to find other, productive virgin land.
Some repeated the operation several times. One old man who had
been born in the Tidewater to parents who had both been
indentured servants ended up in the Ohio Valley, far richer than
he had been in the Piedmont forty years before. He had moved
just 500 miles in his entire life.

Inevitably, the War between England and France (1756–63)
was seen by some Virginians as a war for clean tobacco land. There
were three outcomes of the British triumph. For the Amerindians,
it was a disaster; there was no longer any sort of balance of power
between French and British. For white Virginians, who claimed all
land West to the Pacific, in line with the Colony's northern and
southern boundaries, the opponents of expansion ceased to be the

French and became London politicians. They wanted to limit westward expansion to the Proclamation Line which was, however, rendered redundant within months of its origin in 1763. The resentful French, poorer after defeat, were to prove allies who guaranteed the success of the American War of Independence fifteen years later. Virginia, endowed with a great deal of fresh land seized from Amerindians, produced more tobacco than ever. George Washington may have been one of the most active, practical advocates of westward expansion, the economic incentive being tobacco. And the way he helped the settlement of the West ultimately placed him in the main stream of opposition to the London Government.

Washington is less famous as a farmer or surveyor than as Father of his country. He served Virginia, then became Commander-in-Chief of the Federal Army, then exerted an important influence on the Constitution and was then elected as the first President. He saved the Continental Army from a commander-in-chief from Massachusetts, which in 1775 provided not only much of the Revolutionary Spirit but also many of the soldiers. It is difficult to imagine any leader from Massachusetts with George Washington's qualities or to believe in the success of an army commanded by an opinionated Puritan. Washington was tall and trim yet powerfully built, a man of great natural presence, confident and secure in his aims and opinions. He was able to transform the insubordinate, often politically corrupt Continental Army outside Boston in 1775 into a force capable of defeating the British. Only he could have held the Army together at Valley Forge in the terrible winter of 1777–8, or played the long game until the French joined the American Colonists in 1780.

In later life, Washington's spirit, a deeper and wider version of the same qualities shared with other Virginian landowners, made his Presidency notable for its success at a fraught time when most of Europe was at war. His success in keeping the young country out of the wars was essential to its development. Unlike many heroes in history, he can be seen to be worthy of most of the high esteem in which his memory is held.

Thomas Jefferson was eleven years younger than Washington, and he lived much longer. A tobacco planter from Charlottesville, he became a member of the Virginian Lower House, the House of Burgesses, in 1769, and of the Continental Congress in 1775–6. He drafted the Declaration of Independence in 1776 and was active in Virginia in the War of Independence. Afterwards, he was Ambassador to France and later, Secretary of State and President. A man of extraordinarily wide curiosity and with a depth of intellect greater than that of Washington and other Virginians, he was none the less unable to work out how to gradually free the slaves, an emancipation he much favoured. By the time of his death, fifty years to the day after the Declaration of Independence, the number of black slaves had multiplied by five times in fifty years. Slavery had become, as Jefferson said, 'a firebell in the night'. Despite this failure, his belief in small government, small farms and small businesses has enlightened American politics to this day.[11]

Thirty-five years after Jefferson's death, slave numbers had increased to four million, and Civil War became almost inevitable. Robert E. Lee was at his house, Arlington, which he had inherited from George Washington through his wife, when the Southern States determined to secede. President Lincoln's reaction was to save the Union. This meant military action, and Lincoln offered Lee the generalship of the Union Armies. Lee remained loyal to his State, Virginia, and became the outstanding general of the war, beloved by Southerners and admired by Union leaders. In 1861 he went South and never again saw Arlington, which became a cemetery and a place of pilgrimage for Union veterans and of interment for heroes, including John F. Kennedy, who was buried at Arlington after his assassination in 1963.

It is difficult to see how the Revolutionary War could have reached a successful conclusion without Washington, how the character of the young United States could have emerged without Jefferson, how the war between the States, lasting four years, could have reached its conclusion without Lee. But it is also true that the great early population and wealth of Virginia led to the handsome nature of the Constitution and the Separation of Powers and the establishment of the first truly secular government in the world. In our own time, it was Virginia's generosity of spirit that made the State the first to elect a black governor.

Virginia in turn owes most of its original *ante-bellum* character to the tobacco crop which, before the 1860s, probably did much more good than harm. The original pattern of settlement in Virginia differed in kind, not degree, from the other Colonies. By 1676, 100 years before the Declaration of Independence, there were farmers rich enough to mimic landowners in England, following also their tradition of service to the community. Needy whites were effectively peasants, and the economic conflict between rich and poor was more marked than elsewhere. Nor did conditions in Europe include African slaves or white indentured servants, neither of whom had wages or voting rights, and whose labour helped to make possible the wealth generated by tobacco. Also essential, of course, was the tobacco monopoly granted by the Stuarts and the way that tobacco culture mined centuries of fertility from the soil of Virginia.[12]

III

In the fifty years between the Great Exhibition of 1851 and the death of Queen Victoria in 1901, tobacco was industrialized and the cigarette began to replace snuff, chewing-tobacco, pipe or cigar. In fifty years, the making of cigarettes became one of the most profitable manufactures in the world. In 1851, cigarettes were made on a very small scale – some by the smoker himself, not yet called 'roll-your-own', others by local tobacconists. Although very few women then smoked, most handmade cigarettes for sale were rolled by women as long as hand-manufacture lasted, in some specialized cases until as late as the Second World War.

In 1851, some mature European ladies (but only a few girls) still sniffed snuff, as in the previous century when snuff was widely used by both genders and there was a fashionable ritual, involving hand and face, as did the smoking of cigarettes much later. Snuffing was originally identified with the French, great leaders of fashion in the eighteenth century, less so by the mid-nineteenth century, but in France as late as 1851 more than half the tobacco used was consumed as snuff, while in the United States chewing-tobacco was all-important. In Britain, the pipe was supreme.

Clay pipes in 1851 were still smoked by working women, many

more by men of the same class, if we are to believe contemporary reports of the life and times of what were then called the lower orders. Few women chewed tobacco, but chewing was popular among sailors and others in occupations where smoking a pipe was not possible, on account of fire-risk or inclement weather. But no tobacco consumers in Europe in 1851 used chewing-tobacco as much as the Americans, whose expertise with the spittoon (or cuspidor) was an essential accompaniment to chewing. Most non-chewers – not only women and not only in Europe – thought that tobacco-chewers were risking the health of their stomachs, their digestions and their friendships, since the practice led to rotten teeth and stinking breath. Despite this social disapproval, it is certain that, in 1851, more than half the tobacco consumed by the inhabitants of the United States was chewed.[13]

Nor was lighting a pipe (or even a cigar) at all easy before matches became widely available after 1860, and clay pipes needed much more care and space than did briar pipes which were much sturdier. No one could imagine long – often delicate – clay pipes being carried safely in the pocket; other early nineteenth-century equipment for pipe-smoking was cumbersome in the extreme. So clay pipes were often at least a foot long, and a ponderous mechanical arrangement for striking a flint was needed to make a spark to set alight a cloth or a piece of paper which was then blown till it flared up in a flame robust enough to light the pipe-tobacco. This procedure was not possible out of doors in any but the calmest weather, and very rarely on the deck of a ship. All in all, the cult of smoking, particularly of the cigarette, owed a great deal to the invention of the match. Before matches, it is likely that most pipes and cigars were lit from fires by spills, rather than by freshly ignited inflammable material puffed into life each time someone wanted to light or relight a cigar or a pipe. It was often because of the practical difficulties associated with burning tobacco that non-inflammable chewing or snuffing was so popular before the 1840s when cigarettes were first made in England, not – as has been claimed – in Turkey.

Early hand-made cigarettes were made of tobacco wrapped in tissue paper and most had a mouthpiece attached to one end. The paper was not yet adapted to burn at the same rate as the tobacco.

Nor was the tobacco of the same sort as today, since bright 'Virginia'-type tobacco, flue-cured, so basic for the success of cigarettes, was not developed in any exportable volume until after the War between the States. No suitable machinery to make or pack cigarettes was available until the late 1880s. All these developments were essential to the industrialization of cigarettes.

So cigarettes started out as a hand-crafted artefact, made with unsuitable paper, loosely filled with the wrong sort of tobacco, improperly shredded, sold loose or slackly packed in a kind of unsealed paper envelope, and relatively expensive. Yet they met a significant social need, as the list below proves, giving dates when 50 per cent of (much more) tobacco was consumed in the form of cigarettes, as against virtually nil in 1851:

1920	UK
1923	Japan
1926	China
1939	Austria
1941	USA
1943	France
1951	Sweden
1955	West Germany
1955	Spain
1961	Belgium
1961	Denmark

This list, and much other statistical information on cigarettes, valid or suspect, has been exposed to the kind of earnest analysis happily practised on Madison Avenue and in other habitats frequented by advertising gurus. There are quite a few doctorates on the subject, often involving the changes following the keen adoption of cigarettes by women, and thereafter by ever younger girls. Many ask the question 'Why?', without achieving much in the way of answers.

The modern cigarette story is a scenario unimaginable in 1851, when only a tiny number of cigarettes were smoked in England – their source of origin in the 1840s – and very few at all elsewhere. Today, over a century and two generations later, though tobacco

consumption per head in almost every country may have declined
in total from its peak, the proportion smoked as cigarettes, albeit
increasingly as tipped cigarettes, has also increased.

The great success of cigarettes would not have been possible
without the evolution of flue-cured tobacco. This, like blotting
paper (among other amenities of modern life), was the result of an
accident recognized long afterwards as a great advantage.

By 1839, curing tobacco was more than two centuries old in
Neo-European hands; it was, as today, nearly always performed by
the grower, before the tobacco leaf became saleable. Curing tobacco
is a skilled operation that brings about chemical changes in the
leaf, but in 1839 these were unknown. Curing leaf successfully in
1839 depended on know-how and familiarity and there was little
scientific explanation as to why what happened came about. Leaf
in those days was originally air-cured, which involved a system
somewhat akin to hay-making. But tobacco was not dried in the
field in the way that grass for hay is. On the contrary, tobacco
leaves were hung by the stems, in bundles, in barns built to
encourage the free passage of air. Air-drying tobacco leaf took from
four to eight weeks and was obviously dependent upon a reasonably
low moisture content in the ambient air. This cheap but uncertain
form of drying, during which the leaves should have turned a
ruddy-brown colour, produced uneven results because the outer
leaves of each bundle dried and turned colour earlier than did those
inside. Some conscientious growers, to frustrate this defect, would
take down the bundles of tobacco leaves after a time and sort them
out so that the damper leaves were on the outside, the drier within.

The later system of drying was more positive but more expensive. Leaves were hung up in a barn in basically the same way as
with air-curing, but on the floor of the barn trenches were arranged
in which moist wood smouldered, filling the barn with smoke.
The resulting cured leaf, which was dark brown in colour, was
produced in about half the time taken by air-curing, two instead
of four weeks being typical. One disadvantage of fire-curing (and a
considerable expense) was that someone had to tend the smoulder-
ing logs day and night to make sure that the fire in the trenches
remained alight.

One night in 1839 in North Carolina, a slave who had been

left in charge of a barn full of leaf being fire-cured fell asleep at his post. When he awoke, he found that the fires were out. Instead of trying to relight them with wet wood, he sensibly used half-charred, half-burnt timber from a nearby pile of logs. A lively, smokeless heat was provoked by the burning charcoal and this caused the tobacco to be lighter, more yellow than normal and much brighter.

The slave who had originally been rebuked (and perhaps beaten) for his delinquency was praised when this particular tobacco, the brightest ever seen in North Carolina, made much more than the normal price for brown, fire-cured leaf. After farmers had experimented with pure charcoal for a time, they realized that what mattered was not the charcoal but the smokeless heat it produced. So, by the time of the Civil War, a few new barns had been built with external furnaces and internal pipes or passages which drew the hot air (without any smoke) across the floor of the barns. These were not unlike Roman grain-dryers or Oriental dryers used for tea. Temperature was controlled using louvres for the hot air and ventilators in the sides and roof of the barn. Using almost any wood, not expensive charcoal, a much brighter, milder, more valuable tobacco leaf was produced at a lower cost and more quickly, in about four days instead of from two to four weeks. This wonderful result had three drawbacks, however. First there was the capital cost of the barn; second, much more skill was required from the operator; third, there was the War between the States from 1861 to 1865, and extensive devastation of the infrastructure of the South. The Union Army indulged in wrecking much of value in the tobacco business, including, at the end, burning thousands of tons of leaf in Richmond, Virginia, and other cities in the Tobacco Belt. In the more rural districts in the South, tobacco barns were victims of Northern revenge against Confederate plantation producers.

One further accident was needed to make mass cigarette manufacture a huge growth industry in the United States. This was the discovery in Ohio in 1864 of a 'sport' – a mutant – in a field of the common Red Burley. This plant was not red but pale, nearly an albino, and it became known as Bright Burley. When Bright Burley was flue-cured, a new form of cigarette tobacco was born and it soon came to be an essential ingredient of US, not English, cigarettes. When the importance of cigarettes grew in England, 'Virginia' or 'Bright-cured' tobacco, as developed in North Carolina, came to dominate leaf exports to England. So, by chance, it happens that US and UK cigarette producers were not in competition for the same tobacco leaf, an important unsung element in the development of the international cigarette story.

Ultimately, Virginia tobacco was grown not only in North Carolina and in the Old Dominion, Virginia, but in four other States and later in many other places in the world. If the tobacco is Bright and its seed is descended from North Carolina stock and the harvest is flue-cured, the resulting leaf is Virginia tobacco, wherever in the world it may be grown today. If the process in the field and in the barn is well managed, the leaf is suitable for the sort of cigarette favoured by the British. But leaf-buyers in the UK still prefer American-grown Virginia tobacco, other factors being equal.

This all became common knowledge soon after the South recovered somewhat from war and Reconstruction in the 1870s. Cigarettes suddenly became a boom product. Worldwide, perhaps, only a few hundred thousand at most were made in 1869, all by hand. Twenty years later, in 1889, nearly 5 billion were made worldwide. Twenty years later again, in 1909, about 5 billion were made in one town, Bristol, England, by one firm, Wills, and about thirty times as many worldwide, as far as can be calculated. No consumer artefact has ever grown so quickly in public esteem as did cigarettes, nor become such a valuable industry, a true creator of wealth. Why?

The answer is physiological more than psychological. Smoke from flue-cured tobacco is acidic, not alkaline as from air-cured or fire-cured tobacco. (Aromatic Oriental tobacco used in American cigarettes is sun-dried and originally from Turkey or the Balkans. Turkish smoke, like Bright flue-cured tobacco smoke, is acidic, which is probably why it was adapted to mix with American tobaccos.) The importance is that acidic smoke is much more easily ingested, when inhaled, than alkaline tobacco smoke, and inhaling makes the nicotine more available and probably more addictive. It is easy for today's revisionist historians to claim that the late Victorians who made and marketed cigarettes knew that their product was an addictive drug. But for fifty years after the 1870s few, even Marxists, made that kind of suggestion, and it was not until 1951 that links between cigarette addiction and cancer were objectively recognized. Long before, back in the 1870s, two other problems had to be solved before cigarette sales took off into the wide blue yonder.

First, there was the mass-production of cigarettes. While a cigarette-worker, nearly always a woman, could make an average of 1,200 cigarettes per day they were not, as already explained, very well made. Cigarettes were soon thought to be one of those objects better made by machine, more especially in the 1870s when machines were held in higher esteem than they are – in many places – today. There were a dozen vain efforts to invent a cigarette-making machine before 1880, in France, the United States and England. No result was satisfactory; but the potential value of mechanization increased at the time because leaf prices had fallen after Reconstruction as the production of cigarette-type leaf had expanded. The fall in leaf price emphasized all the other costs in cigarette production. These were manfacture, packing and distribution – all then hand-work – and marketing, which depended on high output and low ex-works prices. Manufacture, packing and distribution would be modernized in the 1880s, with very important results.

The first successful and efficient machine to make cigarettes was patented and offered to cigarette-making firms in 1881; it was called the Bonsack. As it became the progenitor of great industrial, commercial and social events, it is worth describing its action in detail.

A set of wire brushes swept the filler tobacco on to a moving tape made of linen. This tape fed the filler in between two steel bands which compressed the tobacco; this then became a long round tube of a chosen size. Simultaneously, a strip of paper was fed at right-angles from a drum at the side of the machine, and the paper then surrounded the filler. The whole tube was fed into a 'former' that finally shaped the long tube into its desired diameter by bringing the two sides of paper together, ready for bonding. Another part of the machine then pasted and stuck the two edges of the paper together. Finally, a revolving knife, adjusted to produce cigarettes of the required length, made what was designed to be an even, clear cut.

Cigarettes made by the Bonsack machine, compared with those made by hand by skilled operatives using the same tobacco and the same paper, were at once more uniform, smoother and easier to smoke since they 'drew' better; and the machine-made cigarettes were more economical in the use of paper. The disadvantages were that the cigarettes were not as round as the hand-made, the ends were cut off rather irregularly as to length and the cut itself was often fairly jagged. These drawbacks – none of them insoluble – had to be measured against a saving in labour of over 90 per cent. This meant that every consignment of cigarettes could now be invoiced ex-factory at less than 20 per cent of the hand-made price, duty excluded. Profit could be pursued instead of market share, or there could be a happy mix of both.

This very substantial manufacturing advantage was essential for the future success of cigarettes in both England and America. The Bonsack machine could be improved, and this followed in the next decade, as did the first packing machines. Before the development of these, several hand-workers were needed to pack the cigarettes spewing out of each Bonsack machine at the rate of 8,000 per hour in 1881, more than 10,000 per hour in 1884, 20,000 per hour in 1899. But each woman or girl could inspect and pack by hand in fives or tens only about 1,500 an hour – twenty-five a minute – without damaging the cigarettes.

Wills in England and Duke in North Carolina acquired exclusive rights to Bonsack machines in the United Kingdom and the United States respectively. They also pioneered, in their respective countries, packing machines, stiffer packets often containing cigarette cards, big cartons for domestic distribution and tins for export. The enormous increase in productivity achieved by both firms was wholly due to mechanization, and it led to parallel developments.

From modest beginnings in the early 1880s, Wills and Duke came to dominate cigarette manufacture in each of their two countries; cigarettes became the growth product of the decades after 1880, which would have been impossible if they had still been rolled by hand. In Wills's case, sales grew on average by 50 per cent a year for seventeen years after 1883, comparing like with like; Duke did better, but he specialized in the acquisition of his competitors, which accounted for much of his growth. He was later accused by politicians, not without reason, of seeking to create a monopoly and he ultimately fell foul of the Sherman Anti-Trust Act.[14]

Obviously, marketing developments were coincidentally essential, to make the most of the great productivity available after machinery had been put in place. It is a leftist perception that marketing follows the potential of mass-production, because without marketing, mass-production soon becomes over-production. In the case of cigarettes, there is much contrary evidence about which came first: cigarette-making machines or modern methods of marketing.

Economic sophisticates have considered it redundant to ask which came first – branding or mechanized production – regarding this as akin to asking which came first: the chicken or the egg. So it is difficult for some economists to accept that chewing-tobacco, made in a primitive, almost cottage-industry manner, particularly in the South, was almost the first to use branding in a serious way; this was as early as the later 1600s. Anyone who has seen sailors making tobacco fit for chewing knows how easy it is to make the leaf ready for mastication. In some ways, it is no more difficult than for a cow to make grass fit to chew as its cud. The same kind of fermentation is needed before chewing can take place. Given leaf-tobacco, canvas, sugar, water and a few herbs and spices to choice, almost anyone can make tobacco ready for chewing. So it is not surprising that branding was adopted so early to identify the product that manufacturers wanted to sell. But by the 1870s, branding had begun to seem much more modern. Today it is considered essential for selling soap, foodstuffs and commodities like oil products, in the solemn belief that two products, made to the same international specification, can differ if marketed under different names.

But the branding of tobacco products is older than that of soap or ink or chocolate. By the 1840s, in both the United States and England, tobacco for pipes or for chewing was deliberately branded and advertised. Various names were used, often aimed at linking maker and consumer through easily remembered images, though this kind of spinning was not articulated at the time. Typical were the names of various tobaccos (for example, 'Bird's Eye') or names associated with attractive non-tobacco perceptions ('Cherry Ripe' or 'Wedding Cake'). Branding made nationwide recognition possible both in the United States, where nineteenth-century customers might be 1,000 miles away from the manufacturers, and in England, where brand-names were essential for national distribution. Perhaps it was branding that turned household commodities such as ink, mustard, soap, sugar and chocolate into products identified with clear, pleasant and characteristic images and linked to a particular and reliable manufacturer. Perhaps the more commodity-like the product, the more necessary it was to persuade people that it was not a commodity but a particular product of one maker, as with oil products today.

Branding appears to work with any product, and can therefore be called an important element in consumer-capitalism. This is so even if the purchasing public knows (rationally) that there is little or no difference between one packet of white sugar or another, or that two commercial teas are nearly identical or that one washing powder is very nearly the same as another.

In becoming a product far earlier than most other merchandise, tobacco ceased to be a commodity and became identified with its own brand. Chewing-tobacco, which needed the minimum conversion from the leaf stage, was branded far earlier than pipe-tobacco which, like chewing-tobacco, was for a long time virtually a cottage industry, often made up and sold loose by retail tobacconists. Snuff, which consisted of finely ground tobacco, plus additives to taste, hardly needed high technology; it was a product that any tobacconist could manufacture in the back room of his retail shop – or that the snuffer himself or herself could make. Cigarettes, by contrast, needed to be aggressively marketed once mechanization made it possible to produce them so efficiently in such huge numbers. But in fact cigarettes were branded before mechanization, when rows of women and girls would each roll their few hundreds per shift.

Branding was logically followed by advertising and, equally logically, branding had to precede any serious extensive advertising. In both the United Kingdom and the United States, advertising followed the introduction of cigarette machinery, even though branding had been employed two centuries before any machinery was installed. The tobacco industry had recently become capital-intensive and relatively technological, and now had marketing methods to match its new manufacturing potential.

A few figures from the United States tell the story. In 1881, American cigarette output was over 500 million units, largely made by women and girls manually at a rate of about 300,000 cigarettes per worker per year. Five hundred million units implied more than 1,500 female cigarette-makers, nearly half of them in New York. By the end of the decade, in 1889, most cigarettes were made by machine and one woman could, if trained to operate a new machine, produce 6 million cigarettes a year. Duke, the most aggressive of American cigarette-makers, made 834 million ciga-rettes in 1889, while Wills, the leading (but less aggressive) English cigarette manufacturer, made and sold 'only' 55 million units in the same year. But in England, pipe-tobacco, its manufacture now also largely assisted by machine, was still very much more important and valuable than cigarettes, while in the United States in 1889, chewing-tobacco still achieved five times the sales, by value, of the machine-made cigarettes.

It was profit that changed the story; surpluses on cigarettes in the 1890s proved to be about double that of tobacco, whether pipe or chewing, as compared by weight. This may have had more to do with the increased convenience and social acceptability of cigarettes compared with the pipe, which was often treated – like chewing – with the same sort of social disapproval as is smoking (anything) in the more politically correct regions of the world today. For example, cigarettes became part of the stagecraft in theatres in the 1890s, but it is difficult to find many stage appearances of pipes at that date. The most famous literary pipe addict at the time was Sherlock Holmes, who was illustrated monthly in magazine stories, briar (or meerschaum) pipe usually in mouth. He puffed away and measured problems as needing 'one, two or three pipes, my dear Watson'. He even discussed the merits of various tobaccos with his friend. But he was an exceptionally heavy smoker, and he had also used cocaine, a custom, habit or addiction of which Watson did not approve. As for chewing, it was

more and more disdained in polite society if, indeed, it had ever
been acceptable in the better class of theatre, let alone the finer
drawing-rooms. The obvious reason for this disappointment was
that chewing tobacco always involved spitting, and chewing-gum
was seen as a great social advance at about this date, with spitting
no longer wholly obligatory.

IV

The great success of Wills in Britain and the Duke companies in
the United States, both at home and for export, was evident by
1900. It would not have needed a contemporary prophet to forecast
that one day they would clash. Although sales of cigarettes by Wills
and Duke were inconsequential in 1880, by 1900 each had secured
about half the trade in cigarettes in each of their respective home
countries. Wills had done this by application and innovation, Duke
by using his increasingly powerful position to take over rivals,
sometimes by methods not excluding menace. Both, earlier than
other domestic rivals, employed machinery, branding and advertising. Wills's operations were probably more profitable than Duke's
since the latter used his commercial power to increase market share,
sometimes by manipulating, bullying and taking over rivals rather
than by maximizing profits from what he already owned. Wills, in
1899, 1900 and 1901, made annual profits – which were ultimately
to the benefit of fewer than ten family-partner-shareholders – of
£700,000 (over £50 million today). At the time, this was a return
of over 60 per cent on the total capital employed.

Duke bought Ogdens of Liverpool, a much smaller manufacturer than Wills of Bristol and selling only about a quarter as many
cigarettes, and then ran it at a loss, which amounted to over 10 per
cent of turnover for three years. But when the war with Wills was
over, Duke sold Ogdens to the British for nearly three times what
the company had cost him, including Ogden's accumulated losses.

In September 1901, the Wills board of directors had decided to
'form a strong combination with first-class houses' (they meant
British houses) and fight Duke and his British company, Ogden.
Within the same year, Imperial Tobacco was formed, to include

Wills and other tobacco companies considered to be first-class houses. The new company made 70–75 per cent of all the cigarettes produced in Britain. The war with Duke and Ogdens then entered an intensified, sharper phase which did not last long. By September 1902, a year after Wills's seminal board meeting, peace was declared. The American Duke withdrew from the United Kingdom, having sold Ogdens to Imperial Tobacco, which in turn abandoned any ambitions in the United States. A new company, British-American, took over all the world except for the United States and the British Isles and sold overseas every brand of tobacco of any kind made by either Imperial or Duke's American Tobacco. British-American was in turn owned one-third by Imperial, two-thirds by American Tobacco, but it was a British-registered company. Duke's American Tobacco was left owning about 5 per cent of Imperial, the price that had been paid (in shares) for Ogdens. This made the American company the largest shareholder after members of the Wills family.

The Imperial Tobacco Company (of Great Britain and Ireland) was, in 1902, the most valuable and wealthiest company in the United Kingdom by a considerable margin; Imperial also had the highest profit of any British company of the time, about £2 million net (nearly £150 million today) per annum on average before the disruption of 1914–18. The Wills family continued to enjoy more than two-thirds of this bounty.

But the family were not inclined to ostentation and their Dissenting, Congregationalist faith and traditions lived on. There was a strong family belief in the parable of the talents and every young member of the family had to earn his place before he could be considered fit for the company's management. The other side of Wills's Christian belief was concern for the less fortunate. Like Duke, a Methodist, the Congregationalist Wills family was financially responsible for a new university, in Wills's case, in Bristol, England. They were also generous to their native city in many other ways.

In 1901 the Wills family was many times richer than it had been in 1851, with the company's annual profit rising from £5,000 to £750,000 or by 150 times, and the capital value rose by about the same proportion. This growth was achieved by two generations of family management advancing a family-owned firm. Nor had the personal expenditure of the family members increased by anything like 150 times, except in terms of charitable-giving. The Wills company was also a model employer, not only for altruistic reasons; it believed that better-treated workers were also likely to be more productive.

Wills was one of the first large companies in the world to institute profit-sharing for its workers; their scheme started in 1889, with a bonus of 10 per cent of annual earnings at Christmas. Characteristically, Wills workers were urged to save as much of their bonuses as possible, to be added to the non-contributory pension which was to be enjoyed when they retired; there was also a subsidized canteen. Nor were such benefits a substitute for low remuneration, the wages of unskilled Wills tobacco workers being one-and-a-half times the average pay of other unskilled workers in Bristol. Such circumstances were very unusual in late Victorian times, even if they were sourly damned as paternalistic by those unable to think well of any capitalist.

Duke funded a university which bore his name; this was after he had virtually retired. He was viewed rather differently from Wills by his contemporaries, and respected and admired rather than loved (nor did he ever give any impression that he wanted to be loved). By 1907 he was making 70 per cent of the cigarettes produced in the United States. His companies' activities came to include machinery for handling leaf, for making cigarettes and cigars, snuff, pipes and pipe-tobacco, for packaging, for printing and assembling cartons – almost everything, in fact, to do with tobacco except the actual growing and curing of the original crop.

All this vertical activity made a lot of money, but the dominance of Duke's master company, American Tobacco, was inevitably judged to be anti-competitive and in breach of the Sherman Anti-Trust Act. The Supreme Court case was in 1911 and four strong American companies emerged as successors to the one Duke 'monopoly' as did various successor oil companies to Rockefeller's monopolistic Standard Oil in an earlier Anti-Trust case. In the case of tobacco, the four new cigarette companies were American Tobacco, Ligget & Myers, R. J. Reynolds and P. Lorillard; there were also several smaller companies, mostly involved in auxiliary trades like printing and packaging. The scene was set in the United States for a vast expansion of cigarette-smoking, encouraged in fact by competition, to the dismay of campaigners who thought that it was monopoly that encouraged the unhealthy use of tobacco.[15]

But although the US use of tobacco was very great even before 1914, the ultimate largest user would not be the United States but China. The Chinese smoke over 1,500 billion cigarettes every year, in over 1,000 different brands. In the whole world, nearly 7½ million tonnes of tobacco leaf are grown. This harvest is more than 1 million tonnes (18 per cent) greater than that of coffee beans, from which a less addictive consumable is made.

As early in his successful career as 1889, Duke had claimed that he then spent 20 per cent of gross sales income on advertising; in the same year Wills allocated less than 1 per cent of turnover to sales promotion of every kind apart from bonuses to successful salesmen. Twenty years later, in 1909, the Imperial Tobacco Company spent a larger proportion on advertising; but British promotional costs, if on the increase after 1900, never reached American levels. There were more heavens on earth promised by British politicians than by salesmen of tobacco, and the social engineering offered to the electorate in the United Kingdom in the 1900s in the end became as addictive as tobacco.

In the United States in 1913, one of the four Duke successor companies, R. J. Reynolds, started cigarette manufacture, having previously only made chewing-tobacco. The taste for chewing – and the inevitable spitting that went with it – was gradually declining among the more genteel in the United States, especially in urban and suburban areas where women increasingly ruled the roost. By 1913 most women did not appreciate spitting, and if chewing was to be a habit, then the relatively new chickle-gum was to be favoured.

R. J. Reynolds invented a new brand of cigarette and called it Camel. In the first year, 1913–14, only about 1 million units were sold. By mid-1919, the annual sales were more than 20 billion and this represented nearly 40 per cent of all the cigarettes sold in the United States.

The advertising fraternity, based on Madison Avenue in New

York, later drew attention to the direct advertising of Camel cigarettes rather than promotion by means of premiums and prizes. Billboards, newspapers and magazines all carried copy in favour of Camels and the famous slogan 'I'd walk a mile for a Camel' came along in 1921. But there were two other factors which made Camels unique.

One was the actual tobacco mixture. Before Camels, all US cigarettes were made of pure Turkish or pure Virginia or a mixture of the two. R. J. Reynolds's Camel cigarettes were made of a blend of Virginia (Bright) tobacco and the newer Burley. A very small addition of Turkish was more for subtle aroma than for mainstream taste. Even if few, if any, ever walked a mile for a Camel when one of the other two best-selling brands was available, it was always easier to tell the difference between Camel and the other two than the difference between Lucky Strike and Chesterfield. This was proved, in the 1940s, in blindfold tests that were not used by advertising agencies because they were thought to be 'knocking copy'.

The second Camel innovation was the pack size. Before 1913, all cigarettes had been sold in units of five or ten. Camels were only made available in packs of twenty, and other makers later followed Camel's example and the characteristic US cigarette package became general.

Favourable slogans were soon printed on the Camel packages, which did not include the one about walking a mile for a Camel. Presumably, if you had a pack of Camels in your hand, it would have been vain to have printed this particular slogan. (But what if the pack was empty, or only half-full?)

Until the arrival of the fashion for filter-tips in the late 1950s, these three big cigarette-makers together maintained a market-share of more than 60 per cent of cigarettes sold in the United States. Within the trio, Camel remained the leader, even if its share of all cigarettes fell from more than 40 per cent in the early 1920s to about 25 per cent in the early 1950s. As in England, two World Wars boosted cigarette-smoking, but in the United States the use of *all* tobacco remained constant between the wars, although the use of cigarettes multiplied by three times.

In the United Kingdom there was, as already noted, less faith in advertising. Imperial Tobacco spent less on advertising as a proportion of sales than did the Americans – less than 0.1 per cent of cigarette sales in 1910, about 1 per cent of sales in 1920, rising to 2.2 per cent in 1938, the last full year of the inter-war Anglo-German peace.[16]

It was argued that with Imperial Tobacco responsible for 70 per cent of the whole UK market in 1910, and with that market growing steadily if not spectacularly every year, not much advertising was necessary. Within the growth of tobacco sales, there was one old influence and one new one. The old factor was that cigarettes were gaining market share annually at the expense of sales of other forms of tobacco. In the United Kingdom, chewing-tobacco, cigars and snuff were no longer significant for either UK manufacturers or consumers. The three together were worth less than 5 per cent of the UK makers' market share in 1920, compared with more than half for cigarettes. The balance, of course, was the large but diminishing share for pipe-tobacco.

There was little tolerance of chewed tobacco except among sailors (and then not ashore, except in a few lenient pubs in seaports). There was also little manufacture or provision of modern spittoons in the United Kingdom, and twentieth-century British cuspidors are rare enough to be collector's items, if you like that sort of thing. Cigars were largely imported – the better specimens from Cuba – and socially identified with the rich in fact and fable, rarely seen in trams or buses as on the Continent. Snuffing, still to be witnessed in the Law Courts and often used with ritual effect by counsel and judge alike, was no longer considered altogether fit for polite society. Snuff-taking always entails a great deal of post-indulgence sneezing and much work with a handkerchief, which inevitably becomes as soiled with surplus tobacco dust and mucus as a spittoon, and as much to be avoided in the drawing-room.

Tobacco to be burnt in pipes had greatly increased in acceptability since economical, portable and efficient briar pipes became common in the 1880s, but pipe-tobacco smoke was still considered to be stronger than cigarette smoke, and mannish. Pipe-smoking, even though it did not usually entail inhaling, was rarely seen as a twentieth-century practice for middle-class women, although a century before, in working-class towns, smoking a clay pipe was as common among women as wearing clogs. But this was not the case in polite society, though even as late as the 1920s women in some industrial towns could still be found wearing clogs and smoking clay pipes. The pipes were often upside down because it was frequently raining – or at least very damp and foggy – in places such as those that Lowry painted.

Outside the Industrial North, women, if they took up smoking, toyed with cigarettes, and statistics suggest that 'toy' was the right word. Although women like Florence Nightingale (b. 1820), Beatrice Webb (b. 1858) and many other Victorian women smoked cigarettes occasionally, these were probably hand-made of Turkish tobacco, and their use by women before 1920 was almost certainly always in private. Some British women, involved in the First World War as nurses or as war workers, took up serious smoking, but it was estimated that less than 0.5 per cent of all adult women smoked in 1920. This grew inexorably to 5 per cent in 1930, and to 10 per cent in 1939, after which the proportion galloped ahead. In the 1920s smoking cigarettes was seen as an outward symbol of emancipation, starting, as did other symbols of women's liberation, amongst the educated, radical upper-middle class. But only in the 1930s did it become common for actresses to smoke on stage or in films, although Noël Coward, like Oscar Wilde and other theatrical 'progressives', had promoted cigarettes in plays.[17]

Like Coward himself, forced unwillingly in the 1920s to smoke a pipe while playing a character, some non-smoking actresses had to perform reluctantly with a cigarette. To politically-correct people who never today see a smoker on the TV screen or stage, this would now be in breach of the Universal Declaration on Human Rights. For the young, it is one of the wonders of ancient films shown on TV that every adult appears to be smoking much of the time. A half-century ago, one of the games for small minds of every age was to search for continuity errors in films; for example, there were famous occasions when the length of the cigarette smoked by the hero or heroine grew instead of shrinking.

Advertisers targeted women, more especially in the United States, where the use of cigarettes by women (and their general emancipation) grew faster than in the United Kingdom. Lucky Strike was said to be popular among artistes because the smoke did not affect the throat, the tobacco being toasted. Alternatively, to achieve a svelte, slim, modern silhouette, people were advised to 'Reach for a Lucky instead of a Sweet'. Chesterfield represented a man smoking and a woman imploring him to 'blow some smoke my way'. (This, of course, meant the now dreaded passive smoking.) Some copy-writer improved on the ten-year-old 'I'd walk a mile for a Camel' by adding an attractive girl and ' – but a Miss is as good as a mile'.

All the advertising men's highly paid, highly regarded work sometimes absorbed as much as 70 per cent of the net surplus of the three major brands. Even so, the relative positions of Camel, Lucky Strike and Chesterfield were first, second and third in 1920 and this order still held in 1950. The number of cigarettes made and sold by the three brands had increased by four times, but their total market share had declined from 77 per cent to 67 per cent. The chief beneficiary of all the effort was the Lucky Strike brand, whose market share grew 44 per cent in thirty years at the expense of the other two. Those not seduced by advertising will wonder what would have happened if no one had been allowed to advertise any cigarette. It would have meant much higher profits, but opponents of smoking claim that it is advertising that makes people smoke. Yet in the 1920s, cigarette sales failed to grow as much (1.2 per cent annually) in the United Kingdom as they did in the 1930s (6 per cent annually). In the 1930s, because of financial stringency, there was less advertising as a percentage of net profit than in the 1920s, partly because promotion money was spent on gifts, bonuses to retailers, coupons for smokers and cigarette cards, rather than advertising.[18]

*

It would be fair to claim that the major human stress in the First World War was on the Western Front, with its mud, dirt, disease, boredom, fear and danger. For the millions in the trenches, the conditions of daily life, without the hazard of wounds (a clean injury was welcomed as an alternative to death) or extinction from several new, unpleasant weapons, were bad enough to encourage everyone to smoke. The Germans tended to favour cigars and some of the Prussian regiments issued as many as ten to twelve small cigars per man per day during the march through Belgium and France in August 1914. The same ration had proved its worth in the early part of the war of 1870–71. The reasoning was that tobacco restrained appetite and food was more difficult to transport than cigars. But this was the action of an Army Command that at one time considered issuing daily cocaine to every soldier, until it was judged that not enough could be procured to make such a policy possible for several million men. If circumstances had been somewhat different, the German Army of 1914 might have been high on cocaine and the Imperial Army would have performed even greater feats than it did on tobacco. As it was, the soldiers' intoxication was neither from cocaine nor from tobacco, but from extreme patriotism. As the war progressed and the British blockade became more effective, the tobacco available was almost 100 per cent from the Old World – European and Turkish – and there were loud complaints about the resulting quality of both cigars and pipe-tobacco as early as 1915. Cigarettes in wartime Germany were said to be even further from pre-war quality.

In France, the cigarette popular with the troops, the Gauloise, was still made – as in peacetime – of French-grown tobacco with its characteristic aroma – scent, odour or stink, according to taste. It is said that the Germans could tell from the smell of tobacco smoke wafting from the enemy lines which Allied Army was occupying which trenches. The ill-paid, ill-led French Army was supported by plenty of cheap cigarettes and cheap red wine, all in aid of morale. Regrettably, *vin rouge* and Gauloises proved inadequate by 1917 when French Army morale broke and many army units mutinied. This led to a greater strain placed on their British Allies, whose morale was upheld not by red wine but by more and better food and better cigarettes.

There are three stories from 1914–1918 about Wild Woodbines which had been the most popular British brand since 1890, and held about 40 per cent of the UK cigarette market before 1914. It was a small cigarette, selling at the cheapest price, but made of good Virginia tobacco. The first story is that the Prince of Wales (later the Duke of Windsor) always smoked Woodbines at the Front. The second is that tobacco in Woodbines was better than tobacco in more expensive brands. The third involved an

Army padre who never preached but always handed out Wood-bines, earning himself the title 'Woodbine Willie'. The first two stories, if plausible, are almost certainly not 100 per cent true; the quality of the third is that over 1,000 unconnected male mourners attended Woodbine Willie's funeral in 1927, only because they admired him or had met him on the Western Front. His legend lives on, although the cigarette that gave him his nickname is much changed.

Vast quantities of well-made cigarettes were supplied cheap for the troops in the UK and US Services in both World Wars. In the Royal Navy, duty-free tobacco was supplied in leaf, or made up as pipe-tobacco, cigarettes or, occasionally, cigars. Those sailors who prided themselves on their domestic skills drew their tobacco as leaf and prepared it as English sailors one, two or three centur-ies before. Leaf was cut, laced with molasses or rum, and other ingredients added to taste; then it was massaged and pummelled and wrapped in cotton or paper or both, bound tightly with cord and left to 'mature'. Sometimes the maturing roll was left for as long as six months and the result was more or less successful according to the skill of the sailor. If the effect was to produce a tobacco of an offensive strength, the perpetrator could always pretend that this was his intention. Whether his shipmates believed him was another matter.

In the British Army, cigarettes were supplied in quantity to alleviate conditions in the trenches, and many non-smokers changed their habits for the comfort of a good smoke. A recurring legend attached to the Western Front actually originated in the South African War of 1899–1902. Boer farmers were very fine marks-men and they were equipped with good rifles, some of them with telescopic sights. British Tommies relaxed by smoking cigarettes, with which they were liberally supplied. But lighting cigarettes was difficult because of the wind on the Veldt, except at night when the wind fell. One cigarette lit at night was safe; so was a second, but a third often drew a lethal response from a Boer sniper. Several British soldiers were killed before campfire gossip suggested that to light three cigarettes with one match was unlucky. This rumour spread, not only to troops in the First World War but also to civilians, and became a shibboleth for half a century, even if few knew of its true origin in the South African War. In the Second World War, the legend was even diffused to the US forces in Britain and credited to the US Army in 1918.

Ample supplies of all sorts of creature comforts were available for the US forces during and after the Second World War and impressed those Europeans who had suffered from increasing shortages since 1939. Nylon stockings, first sold in 1939 and unknown in Europe before the Americans arrived, were often the price of an hour or two with a good-time girl. American forms of alcohol were new to the ravaged lands and bombed-out cities. But cigarettes, highly valued and in short supply, became a form of currency in some countries, notably in occupied Germany from mid-1945 until the 1948 currency reform.

Hitler had long been a vegetarian and a teetotaller; he was also a rabid anti-smoker, who thought burning tobacco for pleasure more wicked than burning Jews. He ordered the first (reasonably scientific) medical investigation into the causal link between cancer and the smoking of cigarettes, having already banned smoking by uniformed police, SA and SS men in public, even when off-duty. He had ordered severe restrictions on the advertising of cigarettes. He also seriously considered, in 1942, stopping the supply of tobacco to the German Army in Russia. Though dissuaded by the General Staff, he need not have worried: at Stalingrad, at the end of 1942, the supply of cigarettes failed, to the despair of stricken German troops. Some, as prisoners of the Soviets, were marched into captivity in Siberia and did not taste tobacco again for more than twelve years, until they returned home in 1955.

After Hitler committed suicide in the bunker in Berlin in April 1945, as soon as his body had been burnt, his staff began smoking at work – a solace denied them during his lifetime. But had they consistently carried cigarettes to 'the Office'?[19]

Other countries, other *mores*. In every war-stricken nation where cigarettes were still more or less freely available, women started to smoke in earnest. In the United Kingdom, women smoking in 1939 (10 per cent of the female population) multiplied by over three times by 1945. In the United States, cigarette use by women was identified more with elegance, sophistication and sex-appeal, and linked to icons like Marlene Dietrich. This was a very different, near-Hollywood image compared with the drab wartime European women with their not-very-kempt hair covered by a scarf, and with a cigarette hanging from their lips. To be fair, European women in 1945 had suffered shortages for years and would suffer for most of a further decade until their countries returned to something better than the pre-war norm. The material gap between the United States and Europe was wider in 1945 than ever before or since.

In the West, things dramatically improved after the Marshall Plan. In Eastern Europe, Communism would guarantee a half-century of scarcity, failure and material poverty, for which there was little or no spiritual compensation. Communist rule was only maintained by institutional terror; and, amongst the amenities, good cigarettes were in short supply and international brands generally unobtainable until after the fall of Communism in the 1990s.

V

A few years after Hitler's death in the ruins of Berlin, non-Nazi researchers started to publish not always tentative conclusions about the statistical and causal connections between cigarette-smoking and ill-health. Some seem to have had impure motives and the more extreme often appeared to be driven by a deep moral hatred of tobacco and contempt for all who used it. By the end of 1950, there was statistical evidence, of unequal value, from the United Kingdom, Germany, the Netherlands and the United States. Some of the 'evidence' was alarming when transformed by the tabloid press.

In one case, it was reported that a smoker was ten times as much at risk of contracting heart disease within a year as a non-smoker. Careful examination of the figures, not of course possible in the tabloids, revealed that there was a reduction for smokers of from 99.95 per cent to 99.5 per cent in the chance of remaining free of heart disease (over a twelve-month period). The mirror image was that the risk increased from 0.05 per cent to 0.5 per cent. The risk had gone from tiny to very small, but if the figures were correct, it was still true that a smoker was in ten times as much danger of self-inflicted damage as a non-smoker.

The other philosophic problem was the so-called 'Causal Link'. In the case of some 'evidence', the statistical testament could be called coincidental; for example, after 1950 there was less disease associated with cigarette-smoking in rural areas than in towns. But this difference was less marked in countries like the United States where diesel motors were less common than in other countries where they were proportionately more in use. Then in some countries where vehicle-fuel was still rationed and in short supply, there was less cancer of the lung than in other countries where fuel was freely available. In other cases the statistics could be attacked as not proving a *causal* case.

There was an overriding argument that would not really be demolished for a decade. This was that people most inclined to stress were likely to use tobacco. People prone to stress were liable to suffer heart disease or cancer, with or without tobacco. Others claimed that nothing was proved or disproved by this argument and that it was only a sophisticated form of denial. But for anyone who had experienced the strains and alarms of war, and this included millions of European civilians as well as those in the Services, the stresses were clearly relieved by the succour of tobacco. It was coincidentally suggested that the way in which warfare had become increasingly murderous and grim in the 400 years since tobacco had reached Europe could be construed as a 'causal connection' as valid as the one between disease and cigarette-smoking. (The hypothesis was that without tobacco, troops would not have long sustained the strain of modern war.) Or there was the invention of matches, which alone made possible the popularity of cigarettes after 1870 and without which tobacco could never have prospered as it did. These and other arguments were repeated just as, fifty years later, arguments 'proved' that there was no such phenomenon as global warming or, if it did in fact exist, it had little or nothing to do with human activity. Both types of argument were rehearsed at length by a diminishing band.

From 1950, it took about fifteen years for the anti-smoking lobby to win over public opinion. During the 1950s, for example, while TV was entering its profitable adolescence, cigarettes were still commonly flaunted on air. Those who confessed to being regular smokers as a significant proportion of the population did not decline in most of the developed world until the 1960s; even then, the proportion among men fell three to six years earlier than did the same proportion among women. The number of cigarettes smoked by each sex showed a different trend: a fall for many remaining male smokers from 1962, but from 1955 to 1965 a rise for the – fewer – women of as much as 25 per cent in the number of cigarettes they smoked. (The 'average' woman smoked only two-thirds as many cigarettes as the 'average' man in the decade before 1965.)

TV advertising of cigarettes, an important target for UK reformers, increased twenty-fold in the United Kingdom between 1956 (the first full year of commercial TV) and 1965 (the last full year before restrictions came in). In the next ten, non-TV years, cigarette sales grew by more each year than they had during the ten TV years, which tended to confound campaigners. Children, faced with parental and other bans on smoking, naturally adopted cigarettes as a matter of course. 'Behind the bike shed' became a catch-phrase in schools, while some sophisticated adults thought that outlawing something as relatively harmless as tobacco might mean that fewer adolescents would be attracted to soft or hard illegal drugs. But in the 1960s, as cigarette-smoking among adults declined, less socially acceptable addictions amongst the young increased significantly. No one had the gall to claim that there was any real causal or statistical connection here. As always, it was the use or non-use by the icons of youth or by their peers that tended to dictate participation.

In the United Kingdom, plagued until the 1980s by a structural shortage of dollars, tobacco leaf was a restricted import until 1955. The composition of leaf imports had changed from nearly 85 per cent from the United States in 1944–5, largely funded by Lend-Lease and using the shortest ocean route, to dollar-tobacco becoming only just over half the total imported in 1954–5. The tobacco-blender's art became of great value since non-American leaf, though of 'Virginia'-type, was often the only 'genuine' article available. For both UK and US consumers, important changes were in the growth of filter-tipped cigarettes, the 'tar' question and branding. These three developments were interconnected.

In the United States, in the six years between the Peace and the 'causal relationship' disclosure of 1951, the three chief brands – Camel, Chesterfield and Lucky Strike – were unable to raise their individual proportions of cigarettes sold, despite heavy expenditure on advertising and promotion. At the same time, brands outside the mainstream were carving out their own niches, which would later become important. There were mentholated, king-size and filter-tipped cigarettes, each branded and boosted as something quite different from the norm. New brands – nearly 100 of them – were launched between 1945 and 1955. Most were forgotten, but some were potentially successful; this was at a time when the health scare was gathering force in the United States. For the first time in history, health warnings had an effect on numbers of cigarettes sold.

There were two immediate, industry-wide responses. First, there was the creation of relatively respectable 'research' publicity and a lobbying organization whose chief efforts were directed at undoing what were seen as unreliable statistics and partially or wholly flawed research results. The second reaction was practical, to develop a cigarette in which the public could have more confidence. 'Health' became a positive selling-point, increasingly underpinning promotion and propaganda.

Every tobacco company of any size in the United States, which was the leading country in the filter-tip revolution, chose a name, consulted what would be known one day as a 'focus group' and then tried to find a reason why one filter-tip cigarette – their own – was so much better than those made by others. Filters were capable, it was claimed in each case, of filtering out harmful and troublesome constituents in the smoke; every manufacturer claimed that the filter in his cigarettes was superior to any other, the result of years of research, new material and even newer technology. In fact, of course, filters were made of paper, or cellulose made to act like paper, and all that needed to be done was to ensure that the tip drew at the same rate of suction as (or more easily than) the tobacco in the rest of the cigarette.

The bonus for the manufacturer was that filters were cheaper per millimetre than the tobacco in the rest of the cigarette. If this was relevant in the United States, where taxes on tobacco were low, it was even more relevant in the United Kingdom, where tax in 1951 was three times (in real terms) what it had been in 1939, while the cost of untaxed leaf had increased – in dollar terms – very little. In most of Continental Europe where tobacco manufacture was a State enterprise, the filter evolution offered the same rise in profits that US tobacco manufacturers enjoyed.

In 1951 cigarettes represented 80 per cent of tobacco sales by weight in the United Kingdom, and rather less in the United States. Only a tiny proportion of these sales were filtered cigarettes, less than 0.5 per cent in both countries in 1951. The proportion had risen to more than a hundred times as much – to more than half the market – by the time President Kennedy was assassinated in 1963. Filter cigarettes had achieved more than 75 per cent of all sales in November 1972, when Nixon was re-elected, and by 2001 the figure was over 95 per cent. In some places in the United States, in 2001, it was difficult to buy plain, unfiltered cigarettes – called 'men's smokes' – and these tended to be available in only a few places also likely to be still selling chewing-tobacco.

The changeover to filters followed in the United Kingdom with a time-lag of about four years. By 1970, nearly three-quarters of UK cigarette sales were filter-tipped. Japan was ahead, with over 90 per cent filter-tips by 1970, but in the same year Indian filter-tipped cigarette sales were only 10 per cent of the whole and in (then Maoist) China even less. Tobacco manufacturers welcomed the filter revolution, which made cigarettes less high in nicotine and tar, the two substances which had been identified by the men in white coats as even more harmful than the irritating elements allegedly removed by filters. The reason why manufacturers warmed to low-tar, low-nicotine cigarettes was partly because such a policy made 'waste' profitable.

For years the stems, scraps and dust off the floors of sheds where tobacco was handled were regarded as scrap. In the United Kingdom, this scrap was returned to Customs, whose officers weighed it and credited the manufacturer against duty. (The duty had already been paid, of course, on the whole weight of tobacco drawn from bonded warehouses.) What Customs men did with the tobacco scrap is not known, but they were meant to burn it – not in their own personal pipes, of course. Or they released it to make insecticides or rat poison.

In the United States, tobacco was taxed at the local level before sale in a store or supermarket. This was because many Americans could grow tobacco in their backyards, so there were few opportunities to tax tobacco leaf. Whilst there was a certain amount of anti-tobacco legislation, involving (separately) nearly thirty States between 1921 and 1930, tobacco was realistically never morally equated with hard liquor.

(The comparison with moonshine rye or corn whiskey is instructive. Those areas in the United States where tobacco could be traditionally grown on an individual basis were also often the neighbourhoods that voted for 'Local Option' and became dry counties after the war between the States. The demon drink was considered a more damaging social addiction than tobacco, especially in the tobacco-growing States. But to be fair, alcohol's effects on health and relationships were far more obvious before the 1960s.)

Scrap tobacco had two great advantages: the stuff was cheap (before the Second World War, it was unusable). Second, at about the time that means were found to turn the scrap into sheets, it was also conveniently discovered – by spectroscopic analysis – that this sheet-tobacco was lower in nicotine.

By 1958 in the United States, the use of reconstituted sheet (made from very cheap scrap) plus the use of filter-tips had significantly increased the number of cigarettes made from one pound of tobacco leaf. The number rose by 15 per cent by 1960 and by 45 per cent by 1967. The difference was between the standard-size Camel/Chesterfield/Lucky Strike in 1950, where the whole paper tube was filled with tobacco, and the filter-tipped, low-tar, low-nicotine cigarette, in which the amount of reconstituted scrap in sheet form was increasing every year. As the cancer scare in the 1960s drove more and more Americans to smoke 'safe' cigarettes, this in turn led to an increasing demand for low-tar, low-nicotine cigarettes, to be met by the supply of filter-tips filled with a growing proportion of treated tobacco leaf and reconstituted sheet.

Low-tar cigarettes came later. The trend of demand for them grew exponentially in the 1970s. In 1969, less than 1 per cent of all cigarettes were claimed to be low-tar, defined as 15 mg or less; by 1980, low-tar specifications were claimed in over 60 per cent of all cigarettes sold.

Another technical triumph greatly assisted this marketing development. This was to expand dried leaf to something near the same volume as 'green' tobacco leaf as harvested in the field. This meant, in fact, making the treated tobacco less dense (and therefore lighter), achieved by a process known in the trade as 'puffing'. This new technique, plus the use of filter-tips and reconstituted sheet, met the increasing demand for low-tar, low-nicotine cigarettes.

The financially beneficial outcome, from the manufacturer's point of view, was that by 1983, 523 cigarettes could be made from one pound of tobacco, compared with 324 in 1953; this represented a decrease in the leaf content of each cigarette of over 60 per cent. Since 1983, manufacturing efficiency has further increased so that it is anticipated that by 2003 twice as many cigarettes will be made from one pound of tobacco compared with fifty years previously.

The down side to this 'improvement' is that if today's addict has need of tar and nicotine, he or she will have to smoke more cigarettes; good for the manufacturer, expensive for the smoker. But there is evidence that smoking filter-tipped cigarettes today is much more symbolic than an answer to the desire for certain addictive chemicals. From the 1990s, a tobacco addict in need of a sustained near-drip-feed of nicotine could always buy a patch. So far, patches have been found to have no injurious effect apart from obviously supporting (and encouraging?) an addiction, even when they are sold to those trying to give up. But there is no life-style virtue in sticking on a patch and it is a secret, not social, affair. The only indisputable merit of patches could be their use by the elderly, who want to experience the benefits of nicotine use, including delaying the possible onset of Alzheimer's. This can now be attained without a painful death from smoking-related cancer or heart disease, or coughing or spluttering or becoming breathless.[20]

The speed with which fashionable smokers changed to filter-tips and then to low-tar and low-nicotine cigarettes contrasts with earlier centuries during which the opposition to tobacco was unscientific and largely ignored. The reason, of course, is cancer, probably the most feared disease of modern times, and the 'causal' connection with smoking. During the twentieth century, in the United States, death from cancer, as a proportion of all deaths, multiplied by six times, allowing for the increased expectation of life between 1900 and 2000. Cancer sufferers in remission, or even those cured by surgery, drugs or radiotherapy, numbered half as many as those who died, untreated, from the disease. It is likely therefore that cancer victims and ex-sufferers made up nearly 15 per cent of all those dying in the United States in 2000. It is equally likely that if the cigarette scare had not occurred fifty years before, the proportion would have been greater. Contrary to popular belief, cigarette-smoking is also probably responsible for cancers other than those of the chest, throat or mouth.

Even if heart problems account for as many or more deaths as cancer, there are probably two reasons why they are less feared. Control has been more successful, especially in the use of drugs, and death from heart diseases is not always as prolonged, undignified, distasteful and potentially painful as death from cancer. Not all cancers, nor all heart problems, can be blamed on cigarettes, of course, but lung cancer is a choking, bitter, painful end. Nor is emphysema to be preferred, and both are directly connected with cigarettes. One is dealing here not only with the scientific advances that connect cigarettes with ill-health, but also with the effective industrial technology that delivers what cigarette-users seem to want. No one should ignore the acidity of cigarette smoke, if cigarettes are identified as culprits more dangerous than pipes or cigars, both of which generate *alkaline* smoke. There is no evidence that anyone has experimented with making cigarettes out of tobacco producing alkaline smoke, and this is an important gap in the industry's knowledge about cancer in relation to cigarettes.

*

The Marlboro story, as a morality tale, has a beginning, a middle and an end. Marlboro started life as a luxury cigarette introduced by a then unimportant company, Philip Morris, in the 1920s. It plodded along slowly, first as a luxury product and then, as the result of a rebranding ploy with new design, colours, packaging and, naturally, image, it became a luxury woman's cigarette. It was full-size, fully-packed, without a filter of course, and made of the best tobacco.

At the beginning of the health scare Marlboro was temporarily withdrawn from the market, at which point it represented less than 0.5 per cent of the total cigarette market-share in the United States. Then Philip Morris hit the jackpot with their filter-tipped cigarette for women launched in the key year 1968, Virginia Slims. Supported by massive advertising, Virginia Slims got into the top dozen best-sellers fifteen years later in 1983. No one outside Philip Morris knows how much it cost to get there.

Meanwhile, Marlboro had been transformed, given a filter, crush-proof red-and-white colours and the famous cowboy connection. The hard-bitten macho Marlboro Man had a simple life compared with that of those who bought the cigarettes. There was a degree of innocence in the concept, but life was uncomplicated in the days of Cowboys and Indians, and Marlboro Man excited nostalgia for the Old West, where men were men and life, if hard, was at least simple and understandable. For the modern smoker, life was probably the very reverse: complex, intricately interwoven with all those annoying concerns and interests to be found every day in modern times. The train is late, people don't return telephone calls, the smoker's child is in trouble at school, his immediate boss has a bad cold and a filthy temper, but with a cigarette he goes back to a more basic life, out West in Marlboro Country. Rationally, all this is as absurd as Walking a Mile for a Camel. But advertising input is rational, even if its effect on the consumer is not; in 1976, Marlboro became the best-selling brand of cigarette in the United States, and world wide by 1985. Together with Virginia Slims, another beneficiary of the black art, Marlboro made its progenitor Philip Morris the largest manufacturer in the United States and in the world as a multinational. Almost off the graph in 1930, with a market-share of less than 0.5 per cent, only rising to 6–7 per cent in the late 1940s, Philip Morris, uplifted mostly by Marlboro, gained nearly 30 per cent of the US market by 1980. This has always been celebrated by the practitioners as a triumph for advertising. Certainly, in the United States, where tobacco leaf is relatively cheap, what is spent on advertising exceeds every other single production cost. In the year 2000, advertising represented more than $3, on average, for every 1,000 cigarettes sold, or about 6 US cents on every pack of twenty.

In the United Kingdom, where tax is by far the most expensive component in the cost of cigarettes to consumers, the amount spent on advertising is less than half that spent in the United States. Price is naturally more important than in America. Nor, one feels, does the Marlboro Man have such a vital resonance as in the home of the Western. Not so in the Third World, however, where he is identified with the Western myth, the Western ideal and the Western stars of films and TV programmes which are, of course, exported all over the Third World.

One personification of the Marlboro Man was a minor Western action, stunt man and rodeo rider called Wayne McLaren. He smoked a pack and a half daily – thirty cigarettes a day, perhaps Marlboros, perhaps not – for about twenty-five years, but by 1992 he was dying of lung cancer. Ten days before he died he gave a newspaper interview: 'My habit has caught up with me. I've spent the last month of my life in an incubator, and I'm telling you, it's just not worth it.' He was fifty-one years old.

NOTES

1. Syphilis in the early 1500s had the same kind of devastating, then inhibiting, influence on sexual behaviour as did HIV-Aids in the 1980s.

2. The Tropical Andes was the home of the potato, but within 100 years of Spanish settlement, the plant had become an important source of food in Europe and in some of the European colonies.

3. The English (later British) Navigation Acts fostered merchant shipping from 1651 for 200 years. Briefly, imports into the Home Country or into British Colonies had to be carried in British vessels or in ships owned by the country of export. Within thirty years, the British Merchant Navy was larger in tonnage than the next biggest, the Dutch. By 1850, British ships represented more than half the fleets in the whole world and the new supporters of Free Trade were of the opinion that the repeal of the Acts would be safe enough and reduce costs. So the Navigation Acts were repealed in 1850.

4. Tobacco in Lord Sheffield's *Observations* (1770–71) represented two-thirds by value of all the American Colonies' exports to Great Britain.

5. The 'Counterblast' by King James was partly spawned by the realization of how much the import of Spanish tobacco was costing the country's trade. Thirty years later, nearly all tobacco was imported from Virginia, and paid for in the King's coin, even if it was grown from seed of Spanish tobacco. The 'Counterblast' is couched in royal language, not that of Shakespeare: 'That the manifold abuses of this vile custom of tobacco-taking . . .' it begins. More popular in the Colonies was *King James's Bible* (the *Authorized Version* in England) of which the King did not write a line.

6. At this date, Ireland was treated as a colony and only allowed to grow primary products. Free Trade between Ireland, Scotland and England and Wales only dates from 1801, nearly 200 years after King James I.

7. No European had ever met such a hungry crop as tobacco, and the four-course rotation that was virtually the first self-sustaining use of arable land was still 150 years in the future. Tobacco and the potato, both *Solanaceae*, are gross feeders and were beyond the experience of Europeans of the 1600s, the only *Solanacea* then known in Europe being Deadly Nightshade, a plant rarely cultivated except, perhaps, by witches or murderers.

8. An empty land, said good farmers like Thomas Jefferson, led to bad husbandry. He made no comment about wives.

9. As important as the Declaration of Independence in 1776 was the publication, in the same year, of Adam Smith's *Wealth of Nations*. By coincidence or not, the United States is a much stronger believer in Adam Smith than is his homeland, Scotland, or indeed any country in Europe, which is today more and more dedicated to political control and 'social-democracy'.

10. The average daily money-wage for an unskilled man in Britain today is about 160 times what it was in 1763, nearly half of this to be attributed to inflation, more than half to a real increase in wages. The numerate in 1763 would have found this as extraordinary as the inventions of railways, cars, aircraft or TV.

11. Thomas Jefferson's versatility is difficult to credit. Among other tasks he set himself the chore of recording the cost of comestibles in Georgetown (now part of Washington, DC), over a forty-year period.

12. The effects of ample virgin land, with its centuries of stored fertility, on the cost of 'British' tobacco must be compared with the much more expensive European leaf. It was not until the nineteenth century that means were found to introduce extra fertility from guano or phosphate rock and only in the mid-twentieth century were chemical means discovered to counteract the worst pests and diseases. Many of the chemical 'cures' for pests are now no longer legally employed. This imposes pressure on the plant-breeder who is expected to find resistance in new genes.

13. In addition to bad breath and stomach-ache, chewing some tobaccos caused cancer of the mouth, throat and the rest of the alimentary system. Chewing-gum, first made of chickle about 1880, then of synthetics, was much preferred by non-chewers in contrast with chewers.

14. The Sherman Anti-Trust Act, passed in 1890, declared that combinations

and conspiracies tending to monopoly were illegal and would result in imprison-
ment or fines or both. Administered vigorously, it led to the break-up of oil, steel,
railroad and tobacco 'combinations'. Some European economists have ascribed to
this Act the low level of support for socialism in the United States compared with
Europe.

15. It was an illusion that competition in place of monopoly would lead to
lower consumption of 'undesirables' like tobacco and alcohol. The illusion did not
last long in the twentieth century.

16. Imperial Tobacco, like Wills before it, had much more faith in marketing
ploys in place of the same expenditure on advertising. So in Britain, coupons,
cigarette cards and 'special offers' were more common than heavy advertising
expenditure.

17. See Lady Bracknell in *The Importance of Being Earnest*, when told that
the hero smoked; 'Good, every man should have an occupation.'

18. Yet cigarette sales increased faster in the Slump years of the 1930s than
in the 1920s boom. Stress?

19. The Germans fought the war with a lower ingestion of tobacco smoke
than did the British and Americans. The tobacco was also very much cruder, but
by 1947–8 everyone knew the virtues of various Anglo-American brands. These
values were carried into the time when most cigarettes were not actually smoked
but served as currency before the reforms of 1948. For example, French Gauloises,
unsmoked in an unbroken pack, were worth less than the same number of Camels
or Lucky Strikes. The value of (unsmoked) cigarettes made it difficult for the
Allies to smoke in public, if they had any sensitivity.

20. Chewing 'nicotine gum' can produce the same sort of cancers as does the
chewing of tobacco.

Bibliography

Adams, Leon D., *The Wines of America*, New York, 1978
Altschul, Siri von Reis, 'Exploring the Herbariums', *Scientific American*, May 1977
American Tobacco Company, *'Sold American!' The First Fifty Years*, America, 1954
Amerine, Maynard A. and Joslyn, Maynard A., *Table Wines: The Technology of Their Production*, California, 1970
Anderson, A., *An Historical and Chronological Deduction of the Origin of Commerce*, rev. edn, 4 cols, London, 1787–89
Anderson, Edgar, *Plants, Man and Life*, London, 1954
Anderson, S., *The Sailing Ship*, New York, 1947
Andrews, C. M., *The Colonial Period of American History*, 4 vols, New Haven, 1938
Barbour, Violet, *Capitalism in Amsterdam in the 17th Century*, Ann Arbor, 1963
Barlow, Colin, *The Natural Rubber Industry, Its Development, Technology, and Economy in Malaysia*, Kuala Lumpur, 1978
Berlin, Isaiah, *Four Essays on Liberty*, London, 1969
Bever, O., 'Why Do Plants Produce Drugs? What Is Their Function in the Plants?', *Quarterly Journal of Crude Drug Research*, 1970
Biddulph, J., *The Pirates of Malabar*, London, 1907
Bird, Anthony, *Early Motor Cars*, London, 1967
Bloch, Marc, *Feudal Society*, London, 1961
Boswell, James, Augustine Birrell (ed.), *The Life of Samuel Johnson*, 6 vols, London, 1904
Bougainville, L. A. de, *Voyage autour du monde*, Paris, 1771
Boxer, C. R., *The Dutch Seaborne Empire*, London, 1965
Brougham, Henry Peter, *An Inquiry into the Colonial Policy of the European Powers*, Edinburgh, 1803; New York, 1969
Bruce, J., *Annals of the East India Company*, London, 1810
Cambridge Economic History
Cambridge History of England
Campbell, John, *The Spanish Empire in America, by an English Merchant*, London, 1747
Campbell, W., *Formosa Under the Dutch*, London, 1903; New York, 1970
Chambers, J. D. and Murgay, G. E., *The Agricultural Revolution*, London, 1966

Champion, Richard, *Considerations on the Present Situation of Great Britain and the United States*, London, 1784

Chang Kwang-chih, *The Archaeology of Ancient China*, New Haven, 1978

Childe, V. Gordon, *The Dawn of European Civilisation*, 6th edn, rev., London, 1973

Clapham, Sir John, *An Economic History of Modern Britain*, 3 vols, Cambridge, 1926–28

——, *The Economic Development of France and Germany, 1815–1914*, 4th edn, Cambridge, 1966

Clark, Colin, *Population Growth and Land Use*, London, 1967

Cocks & Feret, *Bordeaux et ses vins classés par ordre de mérite*, Bordeaux, 1949

Cook, James, J. C. Beaglehole (ed.), *Journals*, 3 vols, Cambridge, 1966

Dalrymple, A., *An Historical Collection of the Several Voyages and Discoveries in the South Pacific Ocean*, 2 vols, London, 1770–71; New York, 1967

Dampier, William, *A New Voyage Round the World*, 3 vols, London, 1697; New York, 1968

Davis, Ralph, *The Rise of the English Shipping Industry in the Seventeenth and Eighteenth Centuries*, London, 1962; New York, 1963

Deane, Phyllis and Cole, W. A., *British Economic Growth 1688–1959*, Cambridge, 1967

Dodge, Ernest, S., *New England and the South Seas*, Cambridge, Mass., 1965

Drabble, John, *Rubber in Malaya, 1876–1922, The Genesis of the Industry*, Kuala Lumpur, 1983

Drummond, J. C., *The Englishman's Food*, London, 1957

Dunhill, Alfred H., *The Gentle Art of Smoking*, London, 1954

Eysenck, Hans J., *Smoking, Health and Personality*, New York, 1965

Fisher, H. A. L., *A History of Europe*, London, 1936

Fisher, Ronald A., *Smoking: The Cancer Controversy*, London, 1959

Flannagan, Roy C., *The Story of Lucky Strike*, Richmond, 1938

Fox, Maxwell, *The Lorillard Story*, New York, 1947

Franklin, Benjamin, A. M. Smyth (ed.), *Writings*, 10 vols, New York, 1907

Gadille, Rolande, *Le Vignoble de la côte bourguignonne*, Paris, 1967

Galbraith, V. H., *Domesday Book – Its Place in Administrative History*, Oxford, 1974

Gibbon, Edward, *The History of the Decline and Fall of the Roman Empire*, London, 1969

Gipson, J. H., *The British Empire Before the American Revolution*, vol. x, *The Triumphant Empire; Thunder-clouds Gather in the West, 1763–1768*, New York, 1961

Gray, L. C., *The History of Agriculture in the Southern States to 1860*, Washington, DC, 1933

Greenberg, M., *British Trade and the Opening of China*, Cambridge, 1951

Hallgarten, S. F., *German Wines*, London, 1976

Hancock, Thomas, *Personal Narrative of the Origin and Progress of the Caoutchouc or India-Rubber Manufacture in England*, London, 1857

Harlow, V. T., *The Founding of the Second British Empire*, 2 vols, London and New York, 1952 and 1964

Harvard Bibliography of American History
Hawkesworth, J., *Voyages in the Southern Hemisphere*, 3 vols, London, 1773
Historical Statistics of the US, Washington, DC, 1960
Hobhouse, Henry, *Seeds of Change*, London, 1999
Howard, Michael, 'Power at Sea', *Adelphi Papers*, no. 124, London, 1976
——, *War in European History*, Oxford, 1975
Huntington, Ellsworth, *Civilisation and Climate*, New Haven, 1924
Huxley, A., *Plant and Planet*, London, 1978
Jacobson, Bobbie, *The Ladykillers: Why Smoking Is a Feminist Issue*, London, 1981
Klerck, E. S. de, *History of the Netherlands East Indies*, 2 vols, Rotterdam, 1938
Knorr, K. E., *World Rubber and Its Regulation*, Palo Alto, 1945
Koop, C. Everett, *Koop: The Memoirs of America's Family Doctor*, New York, 1991
Labaree, L. W., *Royal Government in America: A Study of the British Colonial System before 1783*, New Haven, 1930
Lichine, Alexis, *The Wines of France*, New York, 1969
Lief, Alfred, *Harvey Firestone, Free Man of Enterprise*, New York, 1951
Loomis, Robert S., 'Agricultural Systems,' *Scientific American*, September 1976
Lynch, John, *The Spanish American Revolutions*, London, 1973
Markham, Clements R., *A History of Peru*, London, 1892
Marwick, Arthur, *Britain in the Century of Total War*, London, 1968
McFadyean, Sir Andrew (ed.), *From Early Days*, Singapore, 1979
Medawar, P. B., *The Hope of Progress*, London, 1974
Middleton, Arthur Pierce, *Tobacco Coast: A Maritime History of the Chesapeake Bay in the Colonial Era*, Baltimore, 1984
Milburn, W., *Oriental Commerce; containing a geographical description of the principal places in the East Indies . . . with their Produce, Manufactures and Trade . . .*, 2 vols, London, 1813
Mitchell, B. R., *European Historical Statistics 1750–1970*, London, 1975
——, with Phyllis Deane, *Abstract of British Historical Statistics*, Cambridge, 1962
Morison, Samuel Eliot, *Christopher Columbus, Mariner*, London, 1956
Namier, Sir Lewis, *Crossroads of Power*, London, 1962
Norwood, Richard, *The Seaman's practice, containing a fundamental Problem in Navigation, experimentally verified, viz: touching the Compass of the Earth and Sea, and the Quantity of a Degree in our English Measure, also to keep a reckoning at Sea for all Sailing, etc. etc.*, London, 1637
O'Reilly, Maurice, *The Goodyear Story*, New York, 1983
Origo, Iris, *The Merchant of Prato*, London, 1957
Orleans, C. Tracy and Slade, John (eds), *Nicotine Addiction: Principles and Management*, New York, 1993
Orwin, C. S., *The Open Fields*, Oxford, 1967
Oxford History of Technology
Parry, J. H., *The Age of Reconnaissance*, London, 1963
——, *The Spanish Seaborne Empire*, London, 1966
——, *Trade and Dominion*, London, 1971
Penning-Rowsell, Edmund, *The Wines of Bordeaux*, London, 1979
Philips, C. H., *The East India Company 1784–1834*, Manchester, 1940

Plato, *The Republic*
Plumb, J. H., *Man Versus Society in Eighteenth Century England*, London, 1969
Pole, J. R., *Political Representation in England and the Origins of the American Republic*, London, 1966
Porritt, Benjamin Dawson, *The Early History of the Rubber Industry*, London, 1931
Read, Herbert, *Anarchy and Order: Essays in Politics*, London, 1954
Ridley, H. N., *The Story of the Rubber Industry*, Singapore, 1911
Robert, Joseph C., *The Story of Tobacco in America*, Chapel Hill, 1949
Rostow, Walt (ed.), *The Economies of 'Take off' into Self-sustained Growth*, Washington DC, 1963
Rousseau, Jean-Jacques, *A Discourse on the Origin of Inequality*, London, 1952
Russell, E. John, *The World of the Soil*, London, 1961
Sadler, D. H., intro: *Man Is Not Lost: A Record of 200 Years of Astronomical Navigation with the Nautical Almanac, 1767–1967*, London, 1968
Saintsbury, George, *Notes on a Cellar-Book*, London, 1978
Sauvy, Alfred, *General Theory of Population*, London, 1969
Scott, Dick, *Winemakers of New Zealand*, Auckland, 1965
Simon, André L., *The History of the Wine Trade in England*, London, 1906
Spruce, R., *Notes of a Botanist on the Amazon and Andes*, London, 1908
Steel, David, *Elements and Practice of Rigging and Seamanship*, 2 vols, London, 1794
——, *Elements and Practice of Naval Architecture*, 2 vols, London, 1805
Tacitus, *Histories*, Kenneth Wellesley (trs.), London, 1964
Tawney, R. H., *Religion and the Rise of Capitalism*, London, 1969
Tench-Cox, *A View of the United States of America*, Philadelphia, 1794
Thomas, Hugh, *Cuba or the Pursuit of Freedom*, London, 1971
——, *An Unfinished History of the World*, London, 1979
——, *The Slave Trade*, London, 1986
Tilley, Nannie Mae, *The R. J. Reynolds Tobacco Company*, Chapel Hill, 1948
Tompkins, Eric, *The History of the Pneumatic Tyre*, London, 1981
Towle, Margaret, *The Ethnobotany of Pre-Columbian Peru*, New York, 1961
Trevelyan, G. M., *English Social History*, London, 1942
Tucker, Josiah, *The True Interest of Britain, set forth in regard to the colonies: and the only means of living in peace and harmony with them*, London, 1774
Turnbull, C. Mary, *A Short History of Malaysia, Singapore and Brunei*, Melbourne, 1980
Wald, Nicholas and Froggatt, Sir Peter, *Nicotine, Smoking and the Low Tar Programme*, New York, 1989
Walpole, Horace, *Memoirs of the Reign of King George the Third*, London, 1845
Washington, George, John C. Fitzpatrick (ed.), *Writings*, 39 vols, Washington, DC, 1931–44
Weber, Max, *The Protestant Ethic and the Spirit of Capitalism*, London, 1930
Weddell, H. A., *Voyage dans le nord de la Bolivia*, Paris, 1853
Woodroffe, J. P. and Smith, Harold Hamel, *The Rubber Industry of the Amazon*, London, 1915

Index